Computational Music Science

Series Editors

Guerino Mazzola
Moreno Andreatta

More information about this series at
http://www.springer.com/series/8349

Damián Keller • Victor Lazzarini •
Marcelo S. Pimenta

Editors

Ubiquitous Music

 Springer

Editors
Damián Keller
Amazon Center for Music Research (NAP)
Federal University of Acre (UFAC)
Rio Branco, Brazil

Victor Lazzarini
Department of Music
Maynooth University
Ireland

Marcelo S. Pimenta
Institute of Informatics
Federal University of Rio Grande
    do Sul (UFRGS)
Porto Alegre, Brazil

ISSN 1868-0305   ISSN 1868-0313 (electronic)
Computational Music Science
ISBN 978-3-319-36500-8   ISBN 978-3-319-11152-0 (eBook)
DOI 10.1007/978-3-319-11152-0
Springer Cham Heidelberg New York Dordrecht London

Printed on acid-free paper

Springer is part of Springer Science+Business Media (www.springer.com)

To Conrado Silva, in memoriam.

This book is dedicated to the composer, acoustics engineer and educator Conrado Silva De Marco (1940–2014). Uruguayan by birth and Latin American by conviction, Conrado helped to give shape to an alternative view of music making. A man of few words, he wasn't fond of inflamed speeches or cocky postures. He taught by example. His political convictions were expressed in deeds, not words. His 40 years of teaching left an imprint on the second generation of technologically savvy Brazilian musicians and educators.

Preface

This book represents a first step in trying to define Ubiquitous Music (UbiMus) as an interdisciplinary research area that lies at the intersection of music, computer science, education, creativity studies, psychology and engineering. The contributions gathered in this volume provide a snapshot of 7 years of community efforts. These efforts include permanent interactions through virtual forums, multi-institutional research projects and publications and yearly scientific and artistic meetings, the Ubiquitous Music Workshops (UbiMus).

Some of the texts included in this volume were originally presented at the III and IV UbiMus Workshops in São Paulo (2012) and Porto Alegre (2013), Brazil. Part of the contents of Chap. 1 were presented at the Congress of the Brazilian National Association of Postgraduate Studies and Research in Music (ANPPOM 2011), held in Uberlândia, MG, Brazil. Chapter 6 is an extended version of a paper delivered at the 2014 Linux Audio Conference in Karlsruhe, Germany, and featured as a lecture at the IV UbiMus.

The book traces a particular path to introduce the reader to UbiMus research, featuring theory, applications and technological development. The prologue and the two introductory chapters lay down the current theoretical principles that underpin the area. Several of these ideas bind together the subsequent contributions, providing a context for the artistic and educational applications presented in the book. The section on applications—comprising Chaps. 3, 4 and 5—furnish three examples of UbiMus projects, spanning creative practice and socially aware educational usage. The last section examines specific support technologies required by UbiMus activities, providing the methodological counterpart of the concepts discussed in the first section of the book.

We expect this work will not only appeal to practitioners of UbiMus but become a good resource for researchers and educators in all the correlate areas engaging creativity and computing. It is our hope that with this volume, the debate on socially grounded views of creative practice will embrace both cutting-edge art and theoretically sound knowledge.

Maynooth, Ireland   Victor Lazzarini
Rio Branco, Brazil   Damián Keller
Porto Alegre, Brazil   Marcelo Pimenta
June 2014

Acknowledgements

First of all, we would like to thank our families for understanding and supporting the extra effort we had to dedicate to creating an international research network from scratch.

The authors would like to thank the funding agencies CNPq and CAPES of the Brazilian government for the partial support for the research in this book. We would also like to acknowledge the help from Science Foundation Ireland's International Strategic Cooperation Award, through the Research Brazil Ireland project, which provided funds for travel and meetings of the Ubiquitous Music Group. Our universities, Universidade Federal do Acre, Universidade Federal do Rio Grande do Sul and Maynooth University, also provided funds for the collaborative activities realised during the last 7 years.

All articles presented here have gone through a rigorous peer-review process. Thus, we would like to extend our gratitude to the III and IV UbiMus Programme Committee for providing their expertise: Álvaro Barbosa (Universidade Católica Portuguesa, Portugal), Adolfo Maia Jr. (Universidade de Campinas, Brazil), Andrew R. Brown (Griffith University, Australia), Eduardo Reck Miranda (Plymouth University, UK), Evandro M. Miletto (Instituto Federal do Rio Grande do Sul, Brazil), Georg Essl (University of Michigan, USA), Ian Oakley (Universidade da Madeira, Portugal), Jônatas Manzolli (Universidade de Campinas, Brazil), José Fornari (Universidade de Campinas, Brazil), Juan P. Bello (New York University, USA), Leandro Costalonga (Universidade Federal do Espírito Santo, Brazil), Luciano Vargas Flores (Universidade Federal do Rio Grande do Sul, Brazil), Marcelo Gimenes (Universidade de Campinas, Brazil), Marcelo Johann (Universidade Federal do Rio Grande do Sul, Brazil), Marcelo Queiroz (Universidade de São Paulo, Brazil), Maria Helena de Lima (Universidade Federal do Rio Grande do Sul, Brazil), Paulo Castagna (Universidade do Estado de São Paulo, Brazil), Nuno Otero (Linnaeus University, Sweden) and Rogério Costa (Universidade de São Paulo, Brazil).

We are deeply indebted to our colleagues in the Ubiquitous Music Group, some of which are authors in this volume, for the invaluable discussions and ideas shared throughout the last 7 years. In particular, our thanks go to Leandro Costalonga,

Nuno Otero, Marcelo Queiroz, Daniel Barreiro, Silvio Ferraz, Patrick McGlynn, Flávio Schiavoni, Carlos Kuhn and Flávio Miranda de Farias. Although they have not been directly involved in writing this book, their contributions to the debate have informed many of the principles contained in the following pages.

Finally, we would like to thank the Computational Music Series editors Guerino Mazzola and Moreno Andreatta for the opportunity to participate in this very special collection of works. And last but not least, we thank our contributors, without whom we would not have been able to provide a wide-ranging picture of what Ubiquitous Music is and what it might become in the near future.

Prologue—Ubiquitous Music: A Manifesto

Marcelo S. Pimenta, Damián Keller, and Victor Lazzarini

Abstract In the last few years, our group has been engaged in a multidisciplinary effort to investigate the creative potential of converging forms of social interaction, mobile and distributed technologies and materially grounded artistic practices. We have proposed the adoption of the term Ubiquitous Music (UbiMus) to define practices that empower participants of musical experiences through socially oriented, creativity-enhancing tools. Ubiquitous Music is thus a new area of research that encompasses ubiquitous computing, computer music and human–computer interaction (HCI), presenting immense challenges for developers and artists. This chapter is a manifesto, highlighting the close relationships between Ubiquitous Music research and everyday creativity. We illustrate our ideas by presenting some prototypes that were developed for everyday mobile devices and which constitute both preliminary results and the test bed of our research. We intend to show how UbiMus activities and applications open up opportunities for musical creation by musicians and untrained participants outside studio settings

1 Introduction

The recent advances in computing technologies and products, which are integrating more and more functionality and constantly changing at an increasing rate, have been a motivating force behind changes in the theory and practice of media arts, demanding broader and challenging approaches. As we move from desktop personal computers, firstly to a multi-platform and distributed Internet, with its web and RIAs, and now to mobile convergence devices and multiuser ubiquitous computing environments, we need to cope with the significance of this media in

M.S. Pimenta (✉)
Institute of Informatics, UFRGS, 91501-970 Porto Alegre, RS, Brazil
e-mail: mpimenta@inf.ufrgs.br

D. Keller
Amazon Center for Music Research - NAP, Federal University of Acre, Rio Branco, Brazil
e-mail: dkeller@ccrma.stanford.edu

Victor Lazzarini
Music Department, Maynooth University, Ireland
e-mail: vlazzarini@nuim.ie

many people's lives. As more advanced portable devices become available, it is becoming increasingly important to have more sophisticated and at the same time more intuitive creativity support tools, particularly allowing users to participate in musical activities, like music creation, performance, experimentation and so on.

1.1   Computer Music and Ubiquitous Computing

The use of computers in music has opened up new possibilities for amateur and professional musicians alike. Research in the field of computer music is directed towards the construction of computer systems supporting not only traditional activities (like music composition, performance, music training and education, signal processing and expansion of traditional music sounds, notation study and music analysis), but also some nonconventional and recent activities like music storage and sharing, information retrieval, and classification of musical data (here including music itself and metadata related to music). Now, combining music and mobile technology promises exciting future developments in a rapidly emerging field, called mobile music [7]. But this new field is still relatively small. There are only a handful of applications available in a market of mobile device software that is comparatively very large. Although the desktop market for music software is huge, the handheld market has not yet experienced the same level of interest. However, devices such as mobile phones, PDAs, MP3 players, smartphones and now pads have already brought music to the ever-changing social and geographic locations of their users, and reshaped their experience of the urban landscape.

Indeed, mobile music technology offers countless new artistic, commercial and sociocultural opportunities for music creation, listening and sharing. How can we push forward the already successful combination of music and mobile technology?

Usually, computer music systems are designed for experienced musicians, requiring previous mastering of specific skills and knowledge of specific musical concepts for better usage. However, since Web 2.0 has turned the passive user into an active producer of content (the word "prosumer" has been adopted for people who are both producers and consumers), we are interested in investigating how to provide support for musical activities in a way they can be carried out even by novices in music, that is, users with little or no previous musical knowledge. Some principles related to the technological support for musical activities for novices are discussed by Miletto et al. [14].

So a simple replication of musician-oriented concepts, interfaces, symbols and features, without a careful analysis of their requirements and world views, could result in tools that would seem useless and unusable to these users. It is very important to accomplish the creation of such tools in continuous partnership and dialogue with potential users, including investigations into their concerns and expectations.

Our research results from the interdisciplinary efforts of the Ubiquitous Music Group [5]. In a recent discussion at the online group forum, a broad definition of Ubiquitous Music (UbiMus, in short) was suggested: "Ubiquitous systems of human agents and material resources that afford musical activities through creativity support tools". This definition suits the interdisciplinary character that this research group seeks for the area.

In practice, Ubiquitous Music is music (or musical activities) supported by ubiquitous computing (or ubicomp) concepts and technology [16, 18]. That is the perspective we take in this book. Considering this point of view, "material resources" and "creativity support tools" are—or may be—those various kinds of stationary and portable computing devices usually integrated into ubicomp systems, and "systems" will generally be, or involve, interactive computing subsystems. Our research adopts everyday consumer mobile communication and information devices—i.e. cell phones, smartphones and PDAs, as well as the medium of the web/internet. We are repurposing those devices and software for use in musical activities, and as interfaces to Ubiquitous Music systems, taking benefit from their distinctive capabilities of portability, mobility, and connectivity and above all from their availability to the average person (including novices in music).

1.2   Contributions

The contributions to this book highlight the close relationships between Ubiquitous Music research and a number of other very interesting research issues. They trace a path that goes from the examination of the theory of UbiMus to current examples of its practices, and concluding in the presentation of some technologies that can support its future development.

In the first chapter, Keller et al. place UbiMus research within the context of creativity theories. It covers all the relevant theoretical background in general creativity frameworks and music-oriented models, introducing the creativity factors, magnitudes and constructs that are applicable to UbiMus research. The chapter also discusses the limitations of several models of creativity that have been laid out within the context of music education and music psychology research. Two particular issues are noted: lack of material grounding and no support for the early stages of creative activity. The text also proposes some key concepts to handle ecologically grounded creative practice and everyday musical phenomena.

Chapter 2 examines creativity-centred design methods, including strategies for interaction, signal processing, planning, prototyping and assessment within the UbiMus context. It shows how these were applied in the development of a prototype for mixing in mobile devices, which was assessed in an exploratory field study. Some implications of this experience are discussed with a view of developing future experiments that can target aspects of creative performance in everyday settings.

The third chapter discusses the Repertoire Remix project, which enables remote audience members to dynamically suggest their musical preferences for musical improvisation sessions via a live web-streaming medium. Through a semantic web interface, which encourages remote participants to collaboratively use "stirring" mouse gestures, the size of graphical bubbles containing composers' names is controlled. Their accumulated weight is then interpreted by musicians to improvise. The chapter documents the first pilot run of the system. It explores challenges in its design, which aims to provide a real-time shared music-style arranging system for live improvisation for network-based performances. Finally, the text also interprets the resulting performance by assessing the participants' mouse gestures collected under the pilot run.

Chapter 4 by Brown et al. reports on their experiences using ubiquitous computing devices to introduce music-based creative activities into an Australian school. It shows how the use of mobile tablet computers running music applications allowed students with a limited musical background to explore creativity in a classroom setting. Such activities were purposely designed to contribute towards personal resilience in the students. An overview of this ongoing project is provided, where its aims, objectives, and utilisation of mobile technologies are discussed. The chapter also explores two theoretical frameworks which informed the research design: the meaningful engagement matrix and personal resilience. A complete report on the activities undertaken to date is provided, including results from questionnaires, interviews, musical outcomes and observation

Zawacki and Johann, in Chap. 5, analyse the possibility of implementing analogue and acoustical processes for music production as remote audio servers accessed by the World Wide Web. This is motivated by the simple idea of making a MIDI-controlled analogue synthesiser available to the general public as a batch process recording service. Ancillary questions arise as what other instruments and audio processes could be implemented as servers, providing quality and cost effective remote access from anywhere, to anyone. Ultimately, the goal is to allow musicians to compose, record and mix their music from simple tools in their computers that can also access very high-quality analogue an real instruments for final sound rendering. The chapter considers all types of instruments, ranging from digitally controlled and old analogue synthesisers, to adapted and custom acoustic and electric instruments, effects audio processing units and analogue mixing consoles. The challenges presented by each different classes of tools are discussed, and the possibilities for these Ubiquitous Music technologies are considered.

Chapter 6, the first of two texts discussing software development technologies for UbiMus, explores two approaches to provide a general-purpose audio programming support for web applications. It provides a review of the current state of web audio development, including some early attempts at using the medium for musical applications. The chapter concentrates first on introducing a JavaScript version of the Csound music programming system, which has been created using the Emscripten compiler. Then, in complement to this, a Native Client implementation of the same system is discussed, running as a sandboxed module under the Chromium and Chrome browser.

The final chapter explores concepts of multi-language programming as a viable means of UbiMus prototyping and deployment. It describes the combination of two high-level audio and music programming systems, Faust and Csound. The latter is presented as an heir to MUSIC N-family of domain-specific languages for music and audio programming, while the former is a purely functional language designed to implement digital audio processing flowcharts. The two systems are combined in such a way that Faust extends Csound, providing facilities for just-in-time compilation of audio processing programs. The chapter shows how multi-language environment can also include a third layer provided by an application framework language, via the Csound application programming interface, to provide a customised user interface to the system.

We illustrate some ideas of our group by presenting some results, prototypes or findings that were developed till now and which constitute both preliminary results and the test bed of our research. The tools and ideas described in this book are the result of an interdisciplinary approach for the design of Ubiquitous Music systems. We intend to show how UbiMus activities, concepts and applications open up opportunities for musical creation by musicians and untrained participants, mainly outside studio settings.

2 Concepts of Ubiquitous Music

Ubiquitous Music is thus a new area of research that encompasses ubiquitous computing, computer music and human–computer interaction (HCI), presenting immense challenges for developers and artists.

Ubiquitous Music systems can be defined as UbiMus environments that support multiple forms of interaction and multiple sound sources in an integrated way. Our work stands at the intersection of computer music, in the form of mobile[7] and network/World Wide Web music[19], with open, participative, nontrivial musical practices such as interactive installations, artistic interventions, ecocomposition and cooperative composition. Although this term has recently appeared in the literature [8], there has not been any attempt to define a workable methodology that contemplates both musical and computational issues raised by these practices. On the technology side, as a minimum, Ubiquitous Music systems should support mobility, cooperation (social interaction) and context awareness. Depending on the types of musical activities, systems can be classified as performance systems (mostly pragmatic activities) or compositional systems (mostly epistemic activities) (see [11]).

The notion of computing technology being everywhere and anywhere—known as ubiquitous computing—has fostered much of the research and development in mobile computing. One of its goals is to make technology disappear or, as Mark Weiser, the founder of the field, eloquently expressed it, "to make computing get out of the way" [18].

Ubiquitous Music systems place further demands on interface design which are hard to satisfy within the instrumental paradigm. When engaged in musical activities with portable devices, users need to have access to the state of the system and the location where the action takes place. Depending on the context, devices may provide sensor or actuator capabilities to the system. This situation demands context awareness mechanisms and location-specific configuration of parameters. Thus, a Ubiquitous Music device may end up not being an instrument after all—a passive object that a musician can play. It may be useful to think of it as an agent in a dynamical system that adapts itself to the musical activity, to the local environment and to the other agents that interact with it.

Ubiquitous computing is still facing several challenges, among which is the issue of transparent user interaction [3]. Mobile devices aren't exactly what Mark Weiser [18] has envisioned as tools for interacting with ubiquitous environments—in fact, he suggested that these very tools should disappear in the environment and that we should compute by interacting directly with it (which is the main argument in favour of tangible interfaces for ubicomp). But since research in the area is still investigating possible solutions for this direct interaction, and since it has also the need to experiment now with solutions to the other issues involved, much of this experimentation is being carried out by using mobile devices as practical alternatives for interfacing with ubiquitous environments.

By choosing everyday consumer mobile devices as Ubiquitous Music interfaces, we also intend to address that main concern for breaking barriers that keep nonmusicians away from musically expressing themselves. Our strategy, for this, relies on:

1. Avoiding dependence on specific, dedicated devices for musical interaction, which would imply both monetary and learning costs, as well as the mere substitution of one barrier (difficulty of access to musical instruments) by another, new one
2. Taking advantage of existing favourable conditions, such as technological convergence and increments in processing power and other device resources, what allows us to reuse those resources for other purposes beyond the original intended ones (i.e. resource/device repurposing).

More specifically, some of the reasons in favour of adopting consumer mobile devices as media/interfaces for Ubiquitous Music include:

1. Access to technology. Even in developing countries, a lot of people have cell phones and can have access to wireless networks, making it an appropriate technology. For example, in Brazil, 85 % of the population has mobile phones, whereas 78 % have TV sets at home.
2. Appropriateness of technology. Again, many desktop-based systems come unstuck on the fact that they need a constant supply of electricity and need to be stored in a cool, secure location. Not only can mobile technology go for several days without being plugged in; their low power drain means they can be recharged from solar cells or clockwork generators.
3. Low cost of everyday devices. Everyday devices like cellular handsets, smartphones and handheld computers are reaching affordable prices that encourage a wider user base.

4. Human capacity training. How much training do people need to use the technology? Mobile handset interfaces are familiar to most users in a way that Windows or Linux are not.
5. Integration into daily routines. Will the technology add an extra burden to the users? Mobile technology is very practical (hands-on technology) and wins here over any desktop solution.

Mobile technology is approaching a new stage, in which the increasing number of young but literate users will play a major role. This user wants and is able to manipulate playful, participative and joyful mobile device-based services. This new context requires new critical qualities for success, like user's sense of control and freedom, engagement and playfulness. In the 1950s, Huizinga coined the term Homo Ludens [9], defining humans as playful creatures and suggesting that the "play element" was crucial to the generation of culture. Play includes acting, games, fun—activities that require human action (maybe performance)—and engagement. We are convinced that "creating one's own music" clearly addresses this basic human need and indeed is rapidly originating its specific cultural shifts.

Recently, the tendency for major increase in processing power, memory capabilities and convergence of technologies in ordinary, everyday, ubiquitous consumer mobile devices (cell phones, smartphones, PDAs) has attracted the interest of several computer music researchers and artists. They saw the potential of combining music and mobile devices—something that is already being called mobile music—in experiments that involve supporting mobility intrinsic to musicians, distributed and instant access to musical information and processing, localisation and context awareness (as in locative audio) and connectivity for musical activities [7]. This includes resulting artistic forms, like music emergent from the cooperation in ad hoc networks or from interactions with the environment [1].

One of our goals is to develop tools which take advantage of these inclusive contexts, providing conditions to novices interested in music to participate in creative activities, in any place and at any moment, using the very conventional mobile communications and computing devices that they already own, and are familiar with.

3 Metaphors and Patterns for Ubiquitous Music

Since we have suggested the adoption of the term Ubiquitous Music as a way to encompass complementary artistic and technological practices, we laid out a conceptual framework which suggests the inclusion of compositional works within the context of software development, and we have defined music making as a set of epistemic and pragmatic activities involving agent–agent (social) and agent–object (natural). Three interaction metaphors were developed to cope with such social and natural interactions:

- The cup metaphor appeared in the context of multimedia performance and installation works, where a "cup"—a passive, unbounded, unobtrusive and empty space (instrumented by means of a sensing system consisting of hidden motion detectors)—becomes an interface that may be freely explored by the participants: the action of the audience triggers lights, slide and video projections and sound.
- The time tagging metaphor, where direct couplings between physical and conceptual sonic operations make possible to define how a set of unordered virtual elements or processes is layered onto a tagged timeline. This metaphor was used to implement mixDroid (see the next section) and tested in the composition Green Canopy: On the Road.
- The spatial tagging metaphor, where virtual elements and/or physical elements are used to create channels of natural interaction, establishing a strong relationship among contextual elements (including physical objects, places, usage situations) and the creation. In other words, when creative activities in ubiquitous context demand kinetics interaction, external physical and material elements can be useful as references for cognitive and creative processes. This metaphor was illustrated preliminarily by the work Harpix Minicomp 1.0, realised as a case study for the ubiquitous system Harpix 1.0.

More information about these metaphors is provided by Keller et al. [10] and Pimenta et al. [15].

3.1 Design Patterns

A general problem of UbiMus is designing for new digital technologies such as ubiquitous computing and everyday mobile devices. When developing for such contexts, we cannot focus on specific user interfaces, due to a presumed device independence. Interaction design for ubicomp and for generic mobile devices has to be done from within higher levels of abstraction. We have found that interaction patterns [2, 17] are a suitable means to address both these issues. Initially, in our project we have considered using patterns to encapsulate and to abstract solutions for specific subproblems of interaction design. This way, we could concentrate on the broader conception of an interactive ubiquitous system to support some musical activity, without having to depend on implementation constraints or target platform specifications. This allows focusing on the higher-level human, social and contextual aspects of interacting with these systems.

In software design, a pattern is a portable high-quality solution to a (usually small) commonly recurring problem [6]. Indeed, patterns allow developers to work faster and produce better code by not duplicating effort. When design patterns were introduced to the software community, they created a lot of interest. As a result, they have now been applied to many other areas. Surprisingly, very little attention has been paid to discuss the adoption of patterns for the design of Computer Music applications.

By analysing the state of the art in mobile music systems and applying our own expertise in computer music, we were able to identify four patterns that do abstract possible forms of musical interaction within UbiMus contexts:

- Pattern #1: Natural interaction/natural behaviour, corresponding to musical interaction which imitates real interaction with a sound-producing object. Thus, all musical gestures that we might regard as "natural" may be explored herein: striking, scrubbing, shaking, plucking, bowing, blowing, etc.
- Pattern #2: Event sequencing, allowing the user to access the timeline of the musical piece and to "schedule" musical events in this timeline, making it possible for him/her to arrange a whole set of events at once. Using this pattern, users interact with music by editing sequences of musical events, which may be applied creatively to any type of musical activity (composition, performance, etc.). It is a straightforward way of implementing the time tagging metaphor.
- Pattern #3: Process control, which allows users to control a process which, in turn, generates the actual musical events or musical material. It solves that important problem in mobile music, which is the repurposing of nonspecific devices: how can we "play" a cell phone with its very limited keyboard, not ergonomically suited to be played like a piano keyboard? It corresponds to the control of parameters from a generative musical algorithm, defining a mapping from the (limited) interaction features of mobile devices, not to musical events, but to a small set of musical process parameters.
- Pattern #4: Mixing, which consists in selecting and triggering multiple sounds or other musical material, so that they play simultaneously. If a sound is triggered while another is still playing, they are mixed and played together.

A more detailed description of these patterns can be found in a work by Flores et al. [4].

3.2 Interaction and Mobile Devices

All of the four proposed interaction patterns address, in different ways, the general problem of "How may humans manipulate music and musical information using everyday (nonspecific) mobile devices?" Thus, in a general collection of patterns or a pattern language for mobile interaction design, these proposed patterns could be classified under a "music manipulation" or "multimedia manipulation" category.

This small, initial set of patterns obviously does not mean to be a thorough taxonomy of musical interaction in general. We are also still on the process of compiling other related pattern sets: for interactions made possible by musical ubiquitous computing environments (i.e. involving cooperation, emergence, location awareness, awareness of contextual sound/music resources, etc.) and for musical interfaces (which instantiate musical interaction patterns). Nevertheless, the four patterns listed here already account for musical interaction in ubiquitous environments when a single mobile device is the user interface, plus they suit designs that need to ensure that music can still be made with a mobile device even with no access to pervasive musical resources (in case those are not available or are unreachable, e.g. due to connectivity limitations).

Inside our multidisciplinary Ubiquitous Music project, the subproblem of designing musical interaction with everyday mobile devices was approached through an exploratory investigation. As a first step, we did a survey on the state of the art in mobile music applications and, based on our computer music expertise, we were able to analyse and identify patterns of frequent solutions for musical interaction that were being adopted in those applications.

In parallel, we did brainstorming sessions to conceive possible musical applications for ordinary mobile devices. We kept as a premise in these sessions that we should consider many different ways of manipulating music with mobile devices, even if it would require some trade-off between functionality and creative ways of overcoming device limitations. These exercises produced ideas that were tried on some exploratory prototypes. But as a second phase, during prototypes creation, we were already associating the types of musical interaction chosen for each one of the prototypes with the interaction patterns that we were starting to define.

Hence, we began a feedback, exploratory process, in which the emerging patterns were being applied in the design of the prototypes and, in turn, the experience of designing these was being used to refine the patterns.

After the construction of these exploratory prototypes, through which we investigate possible musical applications for mobile devices, it is very important to recognise that to validate such ideas is a challenge. Indeed, a difficulty faced by designers of musical tools is the slowness of the validation cycle. Because complete integrated systems are hard to implement and test, development strategies usually isolate aspects of musical activity hindering the compositional results. Musicians' usage of the tools may not correspond to the intended design, and integration of multiple elements may give rise to unforeseen problems. As a partial solution to these hurdles, we have introduced music making within the development cycle [10]. By including musical activities in the design process, the usability requirements of the system can be established through actual usage. Fine-grained technical decisions are done after this first cycle has been completed. So rapid deployment is prioritised over testing on a wide user base.

During the last few years, our research group has been investigating the use of computing technology to support novice-oriented computer-based musical activities. The development of this support has followed an interdisciplinary approach, involving experts in computer music, HCI, computer-supported collaborative work (CSCW), ecocomposition and music education, pointing toward a new field defined as Ubiquitous Music [10]. One of the main motivations of our work is the belief that no formal musical knowledge should be required for participating in music creation experiences.

Part of our work aims at identifying those situations where ordinary consumer mobile devices expand possibilities in the context of Ubiquitous Music. To accomplish such goal, we did a survey on the state of the art in mobile music applications, investigated concepts and mechanisms, evaluated alternatives, carried out experiments to verify viability of development and implemented computational support to Ubiquitous Music in mobile devices. We have as well investigated and defined (after results from some experiments involving prototypes and actual users) the features which this support must present. Among these features, we are especially interested in those related to issues of interaction with mobile devices (concerns from the field of Mobile HCI) and also related to issues of musical interfaces for novices. A simple replication of musician-oriented concepts, interfaces, symbols and features, without a careful analysis of their requirements and world perspectives, could result in tools that would seem useless and unusable to these users. It is very important to conduct the creation of such tools in continuous partnership and dialogue with potential users, including investigations into their concerns and expectations.

An interesting outcome of our studies—fully reported by De Lima et al. [12]— is the observation of different requirements for musicians and naïve users in exploratory activities. Musicians find instrumental metaphors straightforward to use and expressive. Nonmusicians do not necessarily share this view. Interfaces based on traditional musical instruments are not rated as expressive and productive when it comes to exploring the musical possibilities of a tool. This profile-specific characteristic confirms not only the previous claims about the importance of a careful design of user interfaces for musical activities [13] but also the consistency of our assumption in the context of UbiMus: the need to search for new theories, concepts, metaphors and patterns in order to find how to adopt these everyday mobile devices in musical activities.

Conclusions

This prologue has presented some ideas that were tried on the exploratory prototypes of possible musical applications for ordinary mobile devices. We always kept as a premise that we should consider many different ways of manipulating music with mobile devices, even if it would require some trade-offs between functionality and creative ways of overcoming device limitations.

Our objective was not to point out what is the best way to relate musical activities and mobile technology (even if this way is possible or desirable) but to introduce some of our Ubiquitous Music ideas by presenting some exploratory prototypes, which worked as both preliminary results and test bed of our research. Our intention is to help designers to understand some issues related to creative usage of mobile technology and to show how UbiMus activities and applications open up opportunities for musical creation by musicians and untrained participants outside studio settings. Surely, the success and effectiveness of each concept described here are the result of cumulative expertise and insights that we should understand and apply appropriately in order to build Ubiquitous Music systems. Therefore, it is

(continued)

yet a limited excursion into a territory which includes many other possible perspectives and paths to explore.

References

1. Behrendt, F.: From calling a cloud to finding the missing track: Artistic approaches to mobile music. In: 2nd International Workshop on Mobile Music Technology. Vancouver, Canada (2005)
2. Borchers, J.: A Pattern Approach to Interaction Design. Wiley, New York (2001)
3. Costa, C.A., Yamin, A.C., Geyer, C.F.R.: Toward a general software infrastructure for ubiquitous computing. IEEE Pervasive Comput. 7(1), 64–73 (2008)
4. Flores, L.V., Pimenta, M.S., Keller, D.: Patterns for the design of musical interaction with everyday mobile devices. In: Proceedings of the 9th Brazilian Symposium on Human Factors in Computing Systems. Belo Horizonte, Brazil (2010)
5. G-ubimus: Ubiquitous Music Group. http://groups.google.com/group/ubiquitousmusic. (2014)
6. Gamma, E., Helm, R., Johnson, R., Vlissides, J.: Design Patterns: Elements of Reusable Object-Oriented Software. Addison-Wesley Longman Publishing Co., Boston (1995)
7. Gaye, L., Holmquist, L.E.: Mobile music technology: Report on an emerging community. In: NIME '06: Proceedings of the 2006 Conference on New Interfaces for Musical Expression, pp. 22–25 (2006)
8. Holmquist, L.E.: Ubiquitous music. Interactions 12(4), 71 ff. (2005). doi:10.1145/1070960.1071002. http://doi.acm.org/10.1145/1070960.1071002
9. Huizinga, J.: Homo Ludens. A Study of Play-Element in Culture. Routledge & K. Paul, London, Boston and Henley (1949). http://books.google.ie/books?id=ALeXRMGU1CsC
10. Keller, D., Flores, L.V., Pimenta, M.S., Capasso, A., Tinajero, P.: Convergent trends toward ubiquitous music. J. New Music Res. 40(3), 265–276 (2011). doi:10.1080/09298215.2011.594514. http://www.tandfonline.com/doi/abs/10.1080/09298215. 2011.594514
11. Kirsh, D., Maglio, P.P.: On distinguishing epistemic from pragmatic action. Cogn. Sci. 18(4), 513–549 (1994). http://citeseerx.ist.psu.edu/viewdoc/summary?doi=10.1.1.15.4327
12. Lima, M.H., Keller, D., Pimenta, M.S., Lazzarini, V., Miletto, E.M.: Creativity-centred design for ubiquitous musical activities: Two case studies. J. Music Technol. Educ. 5(2), 195–222 (2012). doi:10.1386/jmte.5.2.195_1. http://www.ingentaconnect.com/content/intellect/jmte/2012/00000005/00000002/art00008
    author=Evandro M. Miletto and Luciano V. Flores and Marcelo S. Pimenta, Jérôme Rutily and Leonardo Santagada,
13. Miletto, E.M., Flores, L.V., Pimenta, M.S., Rutily, J., Santagada, L.: Interfaces for musical activities and interfaces for musicians are not the same: The case for codes, a web-based environment for cooperative music prototyping. In: 9th International Conference on Multimodal Interfaces, pp. 201–207. ACM, Nagoya (2007)
14. Miletto, E.M., Pimenta, M.S., Bouchet, F., Sansonnet, J.P., Keller, D.: Principles for music creation by novices in networked music environments. J. New Music Res. 40(3), 205–216 (2011). doi:10.1080/09298215.2011.603832. http://www.tandfonline.com/doi/abs/10.1080/09298215.2011.603832
15. Pimenta, M.S., Flores, L.V., Capasso, A., Tinajero, P., Keller, D.: Ubiquitous music: Concepts and metaphors. In: Proceedings of the XII Brazilian Symposium on Computer Music, pp. 139–150. Recife, Brazil (2009)

16. Satyanarayanan, M.: Pervasive computing: Vision and challenges. IEEE Pers. Commun. **8**, 10–17 (2001)
17. Tidwell, J.: Designing Interfaces: Patterns for Effective Interaction Design. O'Reilly, Sebastol (2005)
18. Weiser, M.: The computer for the 21st century. In: Baecker, R.M., Grudin, J., Buxton, W.A.S., Greenberg, S. (eds.) Human-Computer Interaction, pp. 933–940. Morgan Kaufmann, San Francisco (1995). http://dl.acm.org/citation.cfm?id=212925.213017
19. Wyse, L., Subramanian, S.: The viability of the web browser as a computer music platform. Comput. Music J. **37**(4), 10–23 (2013)

Contents

Contributors

Andrew R. Brown Queensland Conservatorium, Griffith University, Brisbane, QLD, Australia

Edward Costello Music Department, Maynooth University, Ireland

Maria Helena de Lima Federal University of Rio Grande do Sul, Porto Alegre, RS, Brazil

Marcelo de Oliveira Johann UFRGS, Porto Alegre, RS, Brazil

Luciano V. Flores Federal University of Rio Grande do Sul, Porto Alegre, RS, Brazil

Amber Hansen Queensland Conservatorium, Griffith University, Brisbane, QLD, Australia

John ffitch Music Department, Maynooth University, Ireland

Damián Keller Amazon Center for Music Research - NAP, Universidade Federal do Acre - Federal University of Acre, Acre, Brazil

Victor Lazzarini Music Department, Maynooth University, Ireland

Marcelo S. Pimenta Federal University of Rio Grande do Sul, Porto Alegre, RS, Brazil

Alanna Stewart School of Public Health, Griffith University, Brisbane, QLD, Australia

Donald Stewart School of Public Health, Griffith University, Brisbane, QLD, Australia

Joseph Timoney Computer Science Department, Maynooth University, Ireland

Akito van Troyer MIT Media Lab, Cambridge, MA, USA

Steven Yi Music Department, Maynooth University, Ireland

Lucas Fialho Zawacki UFRGS, Porto Alegre, RS, Brazil

Part I
Theory

Chapter 1
Ubimus Through the Lens of Creativity Theories

Damián Keller, Victor Lazzarini, and Marcelo S. Pimenta

Abstract We place Ubiquitous Music research within the context of current creativity theories. Given the ongoing theoretical discussion on the relevance of domain-specific or general approaches to creativity, this chapter covers general creativity frameworks and music-oriented models. First, we introduce the creativity factors, magnitudes and constructs that are applicable to UbiMus research. Then we discuss the limitations of several models of creativity that have been laid out within the context of music education and music psychology research. The analysis points to two caveats: lack of material grounding and no support for the early stages of creative activity. Key concepts are proposed to handle ecologically grounded creative practice and everyday musical phenomena.

1.1 Introduction

This chapter situates Ubiquitous Music (UbiMus) research within the context of current theories of creativity. The motivations are twofold. On one hand, experimental findings have pointed to a mismatch between observed UbiMus phenomena and the theoretical support provided by music-oriented approaches. On the other hand, general creativity theories are difficult to apply to specific design problems. Reducing the gap between creativity theoretical models and experimental constructs may help us to obtain a sharper picture of the limitations of the current approaches to creativity support (more on this in Chap. 2). At the same time, opportunities for innovative design can be identified by observing creative practices (design patterns), by applying cognitive models (design actions) and

D. Keller (✉)
Amazon Center for Music Research - NAP, Federal University of Acre, Rio Branco, Brazil
e-mail: dkeller@ccrma.stanford.edu

V. Lazzarini
Music Department, Maynooth University, Ireland
e-mail: vlazzarini@nuim.ie

M.S. Pimenta
Institute of Informatics, UFRGS, 91501-970 Porto Alegre, RS, Brazil
e-mail: mpimenta@inf.ufrgs.br

© Springer International Publishing Switzerland 2014
D. Keller et al. (eds.), *Ubiquitous Music*, Computational Music Science,
DOI 10.1007/978-3-319-11152-0_1

by uncovering relational properties through experimental work (design qualities). These three strategies can be combined to implement creativity support metaphors, targeting both domain-specific and general creativity factors. The analysis of the current general creativity frameworks indicates that Ubiquitous Music research can find support from recent theoretical developments, highlighting three aspects of creative phenomena: potentials, resources and by-products. A thorough assessment of creative practice entails observing the opportunities and limitations of the given conditions, establishing comparisons between multiple creative strategies. Creative products provide good sources of information on the material output of the creative processes, but their analysis does not throw light on the procedural aspects of creativity. A particularly useful variable for the assessment of creative practice is time. Time-based methods provide detailed information on the dynamics of creative processes. Depending on the available resources and on the temporal investment on the activity, Ubiquitous Musical activity may yield different by-products. These by-products encompass both the stakeholders' intended results and the unintended impact of their actions on the environment. Thus, they provide a good picture of the sustainability of the creative activity.

The second section of this chapter provides a review of eight domain-specific proposals. The models can be grouped according to two characteristics, comprising staged, sparsely connected topologies and iterative, highly connected structures. They also vary in their degree of emphasis given to the material, procedural and contextual dimensions. Models that emphasise the material dimension provide the most direct window to experimental observation. These models are well aligned with the theoretical proposals that focus on the assessment of creative products. Nevertheless, product-based models focus on the evaluation of compositional results, thus avoiding questions regarding how those results were obtained. Alternatively, iterative models usually target ongoing observations of creative actions. For example, Collins [11] proposes the adoption of process-based models to study the strategies used by composers during the act of composition. Given that these studies should be longitudinal and as little invasive as possible, the iterative approach presents several methodological difficulties regarding data collection and analysis. Hence, a careful consideration of the theoretical and methodological implications of each proposal may provide alternative avenues for Ubiquitous Music experimental research.

Two new research fronts have been opened by the Ubiquitous Music initiative. Everyday musical phenomena have been delineated as a viable object of research (Chap. 2; [35,42]). These phenomena can be characterised by their settings and their stakeholders. In contrast with professional creative manifestations, everyday musical creativity does not demand specialised training or restricted settings. Another contribution of the UbiMus approach to creativity research is the introduction of ecologically grounded concepts and methods for creative practice. The expansion of the embedded-embodied approach to cognition, resonating with the recent proposals within general creativity studies, points to ecologically grounded techniques as an emerging framework to tackle interaction design [36, 39] and audio support on a variety of platforms (Chaps. 5–7; [38]). The last two sections address five musical projects through an analysis of their usage of the material resources, providing pointers to variables that can be adopted in ecologically oriented UbiMus experimental work. A summary of all the models' constructs is presented, and their theoretical implications are discussed.

1.2 General Creativity Frameworks

Runco's [46] two-level hierarchy divides creativity theories into two broad groups: (1) those that deal with creative potential and (2) those that focus on creative performance (Table 1.1). Studies that take the former approach strive to identify factors that foster or suppress creativity in individuals and human groups. Thus, the focus is on unfulfilled possibilities rather than on actual creative results. Within the former class of theories, some approaches deal with factors related to personality (person) and to places (also called environmental pressures or press). Theories that focus on potentials open opportunities for research on predictors, i.e. variables that influence creative outcomes. Examples of predictors are: personality traits, environmental pressures (impacting long-term and short-term adaptations) and cognitive and social resources available for creative activities (included within the process factors in Runco's two-level hierarchy). Theories of creative performance focus on unambiguously creative behaviour. Hence, they try to link creativity factors to creative processes and products by studying the outcomes of the creative activity.

Runco's two-tier framework incorporates one of the most influential classifications of the basic elements of creativity, the 4Ps: person, product, process and press [44]. Rhodes [44, 305] proposed that "the word creativity is a noun naming the phenomenon in which a person communicates a new concept (which is the product). Mental activity (or mental process) is implicit in the definition, and of course no one could conceive of a person living or operating in a vacuum, so the term press is also implicit". This classification was later expanded to include creative potentials [46] and persuasion [49]. Proposals that focus on persuasion view creativity as the result of the ability of individuals to influence the direction taken by a creative domain. These proposals stress the social value of the creative products, excluding manifestations that focus on personal or subjective aspects of creativity.

The magnitude or scope of creativity provides another way to classify creative phenomena, separating creative outcomes according to their level of achievement. Following [4, 23], four levels of creative magnitude have been proposed: big-c, pro-c, little-c and mini-c [37, 23] (Table 1.2). Big-c or eminent creativity encom-

Table 1.1 Creativity theories according to Runco [46]	Level 1	c-Potential	c-Performance
	Level 2	Person	Products
		Process	Persuasion
		Process/place	Interactions (among c-factors)

Table 1.2 Creativity
magnitudes: the four-c model
[4, 23]

Label	Magnitude	Description
Big-c	Eminent creativity	Socially acknowledged creative performance
pro-c	Professional creativity	Creative performance that does not necessarily imply widespread recognition
little-c	Everyday creativity	Personal creative performance
mini-c	Subjective creativity	Personal creative potential

passes manifestations that are socially established as paradigmatic examples of creative results. Typical examples are published works of art and scientific theories. Eminent creativity manifestations necessarily target creative products; thus, professional products that do not involve wide social exposure are excluded from this class. Personal experiences that lead to creative products are treated within the context of everyday or little-c creativity studies [45]. The same authors provide a finer distinction among everyday creative phenomena by introducing the mini-c category. Mini-c creativity involves internal, subjective and emotional aspects of everyday creativity. Thus, the little-c label is reserved for everyday creative phenomena that result in creative products. Between little-c and big-c phenomena, Kaufman and Beghetto [23] suggest a third type of creative behaviour: professional creativity or pro-c. Pro-c studies focus on professional creative achievements that do not attain eminence.

An alternative to the 4Ps classification has recently been proposed by Glăveanu [19]. The 5As framework features a systemic relationship among three stakeholders: actors, audiences and artefacts (Table 1.3). For Glăveanu, creativity is concerned with the action of an actor or a group of actors, interacting with other actors (audiences) and with the material world. This creative exchange is done through affordances leading to the generation of new and useful artefacts, which feature modified affordances. While actors engage in creative actions, they produce artefacts which are shared with an audience. These actions are done in cultural and social contexts which foster and constrain the creative behaviours. A comparison of the 5As with the 4Ps framework indicates a stronger emphasis on the social and material context of the creative act (Table 1.3). While *person* stands for the individual's characteristics, *actor* stresses the factors that connect the individual with her social context. *Process* focuses on the cognitive mechanisms involved during creative activities. Contrastingly, *action* implies a tight integration of cognition and behaviour, pointing to an embedded-embodied view of cognition [18]. *Product*—which describes the features and the consensus built around creative results—is replaced by *artefact*, encompassing both the material and the behavioural results of creative activity within a specific cultural context. *Place* or *press* defines a set of external variables that condition creative activity. This factor is replaced by *audiences* and *affordances*, underlining the interdependence of the stakeholders' actions with their social and material world.

Table 1.3 Glăveanu's 5As sociocultural proposal and the 4Ps framework (after [19]).

4Ps	Factors description	5As	Factors description
Person	Internal attributes of the person	Actor	Personal attributes in relation to the social context
Process	Cognitive mechanisms	Action	Coordinated psychological and behavioural manifestations
Product	Features of physical objects or the consensus around them	Artefact	Cultural context of artefact production and evaluation
Place/press	The social as an external set of variables	Audience	The interdependence between the creators and the social context
		Affordances	The interdependence between the creators and the material world

Glăveanu's [19] framework is closely aligned with the concepts that have emerged from 17 years of ecologically grounded creative practice in music [3, 7, 8, 12, 24, 25, 29]. Barreiro and Keller [2] proposed musical activity as a new focus for creative experimentation. This higher temporal layer complements the work carried out in ecological modelling, involving the synthesis of sonic events and textures [24, 28, 31]. The core of this approach is the ongoing interaction among agents and objects [24], featuring both natural and social affordances as shapers of opportunities and constraints for creative activity [33]. Ecologically grounded creative practice furnished mechanisms to change the role of the audience from a passive spectator to a co-creator in several artistic endeavours [3, 32, 36]. This practice-based research constitutes one of the three methodological approaches that have been used to deal with the uncharted territory of Ubiquitous Music creativity. Section 1.4 will discuss the relationship between the local settings and the creative activity undertaken in five musical projects. The results of the analysis indicate a set of variables that can be applied in the experimental study of UbiMus phenomena. But before discussing the implications of the eco-cognitive approach, let us deal with some of the recent music-oriented theoretical proposals that have addressed various aspects of creativity research.

1.3   Domain-Specific Creativity Models

Keller et al. [34] reviewed eight music creativity models and discussed how they relate to three creativity dimensions: materials, procedures and context. Through a structural analysis of the proposals, they identified two methodological shortcomings in music-specific applications: *early domain restriction* and *lack of material and social grounding*. Their findings pointed towards a convergence between a theoretical construct labelled compositional paradigm shift and recent creative practices, emphasising the relationship between the availability of resources and the creative activity.

Musical creativity models have been heavily influenced by an early model of creativity. Back in 1926, Wallas proposed that creative processes occur in four stages: *preparation, incubation, illumination* and *validation*. For Wallas, the cognitive dimension is at the core of creation. *Incubation* involves time spent away of the creative activity, and *illumination*—or insight—is the mental process through which ideas coalesce. As we will see, both the sequential-stage topology and the overdue stress on disembodied cognition have left their imprint on several current models.

Bennett's model [5] modifies Wallas's stage structure by separating *sketching, elaboration plus refinement* and *revision* as independent stages (see Fig. 1.1). The material dimension is featured as a key aspect of the model. Compositional activity is triggered by a *germinal idea*. This idea is expanded into a first draft of the work. At this point of the compositional process, Bennett suggests that an iterative cycle is established in which the composer revises the initial idea, produces a new sketch and modifies the first draft. The next stage combines elaboration and refinement of materials leading towards the *final draft* of the work. After the final draft is produced, revisions may still be undertaken.

An interesting feature of Bennett's model is that backtracking is considered an integral part of music creation. The iteration between the first three stages and the possibility of introducing revisions after the work highlight two organisational mechanisms: (1) *convergence*, in which the composer applies successive refinements to the same musical materials, and (2) *divergence*, in which the composer produces new materials which are incorporated as part of the compositional paradigm [20]. Nevertheless, a third possibility—the breakdown of the existing organisational paradigm—demanding a switch to a different set of materials and structural relationships is not contemplated by this model.

Backtracking and iteration are structurally featured in Dingwall's model [13] (Fig. 1.2). The model deals almost exclusively with the material dimension. Cognitive aspects are included through the factors *inspirations and parameters*. The compositional process itself is summarised by three interconnected stages: *genera-*

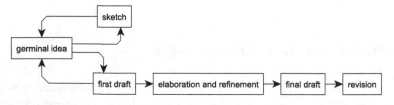

Fig. 1.1 Bennett's [5] music creativity model

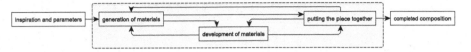

Fig. 1.2 Dingwall's [13] music creativity model

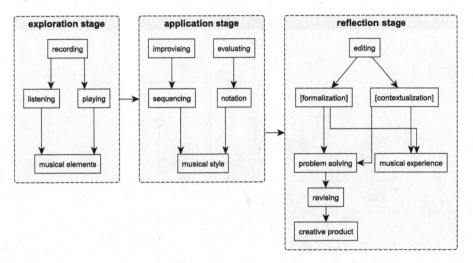

Fig. 1.3 Chen's [10] music creativity model

tion, development [of materials] and *putting [the piece] together.* The result of the process is a *completed composition.*

Webster [52], Chen [10] and Collins [11] emphasise the cognitive dimensions of music composing. Chen lays out a three-stage model that resembles Wallas's proposal (Fig. 1.3). *Exploration* encompasses activities such as recording, playing and listening to materials. *Application* involves synchronous musical decisions—improvising—and asynchronous evaluation of materials (evaluating). The last stage—*reflection*—gathers diverse activities such as editing, problem-solving and revising. Chen also includes publishing as part of the *reflection* stage.

In Webster's model, the material dimension is represented by the results of the compositional activity: the *creative products* (Fig. 1.4). These include scores and recordings of composed music, recorded performances, recorded improvisations and written analyses. Webster also mentions the *mental representation of the music heard* as a type of creative product. Given that this construct can only be observed indirectly, it might be better placed within the cognitive dimensions of the model, i.e. the *thinking process.*

The compositional process initiates by defining a *product intention.* This stage encompasses five possible activities; therefore, this part of Webster's model deals with procedural aspects of the creative process. The composer may choose to compose, perform music of others, listen repeatedly, listen once or improvise.

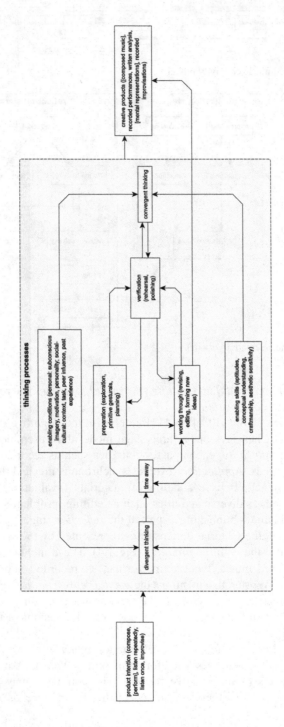

Fig. 1.4 Webster's [52] music creativity model

These five intentions have five corresponding outcomes—the four *creative products* mentioned before plus the *mental representation*.

The core of Webster's model deals with the cognitive dimension, labelled *thinking process*. Three of the four stages of the model follow Wallas' proposal: preparation, verification and incubation. In Webster's model incubation becomes *time away*, and a new stage is added, *working through*. *Working through* involves forming, editing and revising ideas; thus, what were three separate stages in Bennett's model have been compacted into a single stage. The topology indicates an iterative cycle between the four stages with the possibility of going from *preparation* to *working through* without passing through the intermediate stages *time away* or *verification*. The four factors that drive the cognitive dimension are: enabling skills, enabling conditions, convergent thinking and divergent thinking. *Enabling skills* are characteristics of the individual, and *enabling conditions* encompass both personal and sociocultural factors. It is not clear why social aspects were included as part of the cognitive dimension instead of being a separate contextual category. In any case, for Webster the influence of the enabling conditions on the creative process is only indirect: social and personal factors foster *convergent thinking* and *divergent thinking* which, in turn, shape the creative act.

Taking the hint from Shah et al. [48], Collins [11] suggests that compositional activity can be equated to problem-solving activity. From this perspective, composing requires defining a problem and finding solutions for it within a solution space. In Collins' model (Fig. 1.5), the sequence of *solution spaces* involved in creative activity is driven by four cognitive actions: *problem proliferation, finding solutions, deferring solutions* and *restructuring*. Following closely Bennett's *germinal ideas*, the material dimension is represented by *general ideas/themes/motives* which are carried into the *solution space* by defining *subgoals*. Within each solution space, the creative procedure is represented by three sequential processes: *postulating broad aims* (this leads to new ideas/themes/motifs), *small-scale editing* (which yields general solutions or specific solutions) and *see[ing] the broader picture* (this stage conducts to restructuring the problem space). After each iteration, the outcome may consist of *general solutions* or *specific solutions*. If no solution is found, the material is simply taken to the next iteration: the solution is *deferred*. Finding a solution drives the cognitive process towards a new subgoal which may consist of *combined solutions* (involving the *and* operator) or *alternative solutions* (applying

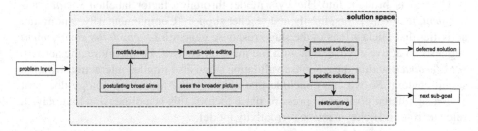

Fig. 1.5 Collins' [11] music creativity model

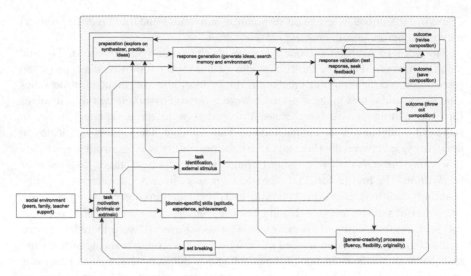

Fig. 1.6 Hickey's [22] creativity model

the *or* operator). Although backtracking is not explicitly specified in Collins' model topology, it is implied that previous solutions may be revisited or that they may be incorporated within the context of new solution spaces.

Hickey's model [22] encompasses five ordered stages which can be reiterated (see Fig. 1.6). The procedural dimension includes *task identification*, *preparation*, *response generation* and *response validation*. The final stage is the *outcome* which features a decision fork: the composer finishes the creative process (*saves* or *throws [the composition away]*) or returns to one of the previous stages (revise [the work]). Cognitive factors are clearly depicted as separate from the creative procedures. Previous creativity research is accounted for by identifying *task motivation*, *domain-relevant skills* and *creativity-relevant processes* as independent constructs. The material dimension is fused with the procedural dimension through references to musical *ideas* and *composition* within the stages preparation, response generation and outcome. In contrast with Bennett's model, Hickey indicates the possibility of a change in compositional paradigm by linking task motivation with creativity-relevant processes through *set-breaking*.

Context is brought into Hickey's model through a factor labelled *social environment*, featuring peers, family and teacher support. Complementarily, among the activities that Hickey lists in the stage response generation, *search the environment* suggests an active engagement with the material context. Furthermore, the activity *seek feedback* within the response validation stage also implies interaction with the social context. So we can conclude that although the model's topology does not give a significant position to press or place factors, this is a dimension that plays a relevant role in Hickey's musical creativity model.

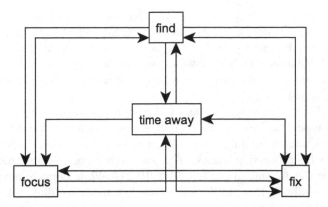

Fig. 1.7 Burnard's and Younker's [6] creativity model

Burnard's and Younker's [6] model is the most streamlined of the eight domain-specific models considered (Fig. 1.7). The authors propose an iterative—*cyclical* in their terms—topology comprising four activities: *find, focus, fix* and *time away*. Thus, the model is strictly procedural. Arrows indicate that activities are not rigidly ordered. All activities may lead to *time away* and vice versa. The activities find, focus and fix may follow any sequential order.

1.3.1 Summary and Implications of Domain-Specific Models for Ubimus Research

Musical creativity models can be grouped according to their emphasis on intrinsic and extrinsic factors. These factors determine the methodology adopted when applying the models in the field, so an understanding of the theoretical background throws light onto the relationships between the model and its applications. Three dimensions can be extracted from the eight models discussed: (1) material (what), corresponding to the resources and the products; (2) procedural (how), encompassing the creative activities; and (3) contextual (where/when), featuring the extrinsic factors.

Regarding the material resources, Bennett's [5] model suggests that compositional processes start from a single germinal idea. Collins [11] also adopts this view but allows for several musical ideas (named themes or motifs) at the initial stage. Contrastingly, Hickey [22], Burnard and Younker [6], Chen [10] and Dingwall [13] models suggest that exploratory activities precede the selection of materials. When sound sources or tools used to generate creative musical products are explicitly considered in the model, the connection to material resources is direct. Two of the three interrelated stages suggested by Dingwall [13]—the generation stage and the development stage—can easily be assessed by measuring the quantity of the material produced. The stage *putting the piece together* may involve selection, grouping and disposal of material resources; therefore, both objective and subjective assessments may be applied. Subjective assessment of creative products can be done through Amabile's [1] Consensual Assessment Technique (CAT). Objective assessment demands measurements of resource yield and resource consumption as a function of time [15].

Procedurally, the reviewed models follow two patterns: iterative and staged. Stage-based topologies rest on the premise that the cognitive dimension is based on serially organised modular cognitive processes. This premise finds support in the symbolic processing views of cognition but goes against the grain of recent approaches to general creativity [17, 19, 21]. While iterative models—such as Hickey's, Collins', Dingwall's and Burnard–Younker's—allow for permanent reassessment of criteria throughout the creative cycle, stage-based topologies imply nonconcurrent access to resources. In other words, decisions at each stage depend only on the resources provided by the previous stage. The topology of Hickey's and Collins' models allows for alternative paths that depend on the previous creative activity. Material resources become available or are discarded during *restructuring* (in Collins' model) or *revising* (in the *outcome* phase of Hickey's model). In Hickey's model, a similar mechanism can be applied to the behavioural resources through the *set-breaking* procedure that connects motivations with general creativity factors.

In more general terms, we can describe the flow of resources using one topological descriptor: connectivity. Highly connected topologies—such as Burnard-Younker's proposal—imply a multiplicity of possible procedures which are very hard to assess in experimental settings. Sparsely connected topologies—exemplified by the stage-based models—provide good opportunities to isolate variables, but the ecological validity of these observations can be questioned from two perspectives. First is the lack of material grounding. Webster's *enabling conditions* include context, task, peer influence and past experience. But these factors are only linked to *convergent thinking*. Thus, they are only relevant for thought processes. Hickey's model accounts for social factors, such as peers, family and teacher support (the *social environment*). Nevertheless, according to the topology of this model, these factors are only relevant to *task motivation*, implying that the environment only influences the creative act through the cognitive factors that drive *task identification* and *response generation*. Second is the early domain restriction. Webster's and Collins' focus on problem-solving may hinder the exploration of activities that have no predefined goals. Creative practice experimental work indicates that the explicit formulation of a procedural target may only occur at late stages of the creative cycle [39, 55]. Therefore, while some pro-c activities may establish clear targets from the start, creative activities by non-specialists and activities that do not adopt instrumentally centred approaches generally demand a long period of exploratory activity before defining the explicit creative goals [36]. The underlying hypothesis is—as suggested by Hickey's, Burnard's and Younker's, Chen's and Dingwall's models—that both restricting and providing access to materials are part of the compositional process. By selecting materials or tools, the experimenter is taking the place of the composer. Hence, the resulting data cannot be used to determine whether the creative musical activity begins by exploratory actions or by a given set of conditions. If the musical materials are limited by the experimenter, it is not possible to draw any conclusions regarding the initial handling of the material resources.

The contextual dimension of musical creativity models encompasses the material and social factors that influence the compositional processes. Material (or physical) factors can be related to two variables: time and place. Most of the reviewed models live in abstract space-time. Material context is generally included as an ad hoc factor. An alternative approach, proposing a tight relationship between the settings where the creative activity takes place and the decision-making process, has emerged from ecologically grounded creative practices. The next section provides an analysis of five cases, targeting a procedural strategy that could deal with time, place and creativity within the context of Ubiquitous Music making.

1.4 Ecologically Grounded Creative Practice

A framework for the study of relationships among creative settings and creative products has emerged from ecologically grounded creative practices [27]. Sonic ecologies—habitats where agents and objects interact producing creative sonic by-products—were proposed as a methodological framework to ground creative musical work [27]. Sonic ecologies encompass at least three dimensions: material resources, human agents and the context where the creative activity takes place. Five music projects serve as contrasting examples of usage of material resources: *It's Gonna Rain* [43], *Kits Beach Soundwalk* [54], *Metrophonie* and *The Urban Corridor* [9, 26] and *Net_Dérive* [50]. The analysis of the five creative projects indicates three operations at work: constrain, expand and shift. Two strategies emerge from the relationships established between the local settings—more broadly defined as the ecological niche (eco-niche)—and the decision-making process that impacts the choice of material resources for creative activities. One strategy targets the use of fixed material resources. The other approach establishes a dynamic relationship between the contextual dimensions and the material resources.

The *shift* operation entails a mismatch between the local resources and the creative products. No matter where the sound sources come from, the musical products are created and delivered within an entirely new context and situation, i.e. the composer's eco-niche does not correspond to the listener's eco-niche. When the shift operation is employed, the material resources do not match the material context of the work. This is the case with most electroacoustic works for fixed medium. *Kits Beach Soundwalk* [54] provides an example of the use of the shift operation.

The *constrain* operation can be observed at work in the piece *It's Gonna Rain* [43]. When comparing *Kits Beach Soundwalk's* attempt at reconstructing the original soundscape of Kitsilano Beach with Reich's selection of a single sample as source material, we see opposite creative strategies are at play. *It's Gonna*

Rain constrains the material dimension from a complex set of relationships of the local soundscape to a single voice sample. The eco-niche is reduced to the sonic interplay of the collected sample with itself. The interesting aspect of the piece lies in its procedural dimension. While *Kits Beach Soundwalk* sticks to the composer-centred creative mechanics—the dynamics of the compositional system are explicitly constructed by the artist—*It's Gonna Rain* takes advantage of an autonomous perceptual phenomenon: emergence. In Reich's piece, the creative decisions at a local level are a consequence of the global mechanism previously defined by the composer: the phase difference among the tracks.

Metrophonie provides an example of the use of a basic organising concept and a concurrent synthesis technique: *accumulation* [28] and *ecological modelling* [31]. First, short samples—or grains—are extracted from the recorded sources. These grains provide the basic spectral and micro-temporal features of the sounds to be synthesised. Short events—modelled after the characteristics of a class of recorded sound events—are synthesised through modelled granular distributions [24]. These models generate sound classes with dynamic meso-temporal characteristics yielding events which are perceptually related but never identical to each other [25]. Thus, *Metrophonie* features a continuum from the recorded material provided by the eco-niche to the synthetic events used as sound sources. In more general terms, the eco-niche is extended along materially meaningful dimensions. The synthetic sound classes are just complements of the recorded sounds, so there is a partial match between the local eco-niche and the composed material. This operation can be defined as *expansion* of the material resources.

While *Metrophonie* proposes the expansion of the material domain, *It's Gonna Rain* provides an example of a constrained eco-niche. Through transparent editing and mixing, *Kits Beach Soundwalk* shows how a shift operation can be applied to the material dimension. If we label the original state of the material resources as S_0 and the final state as S_1, corresponding to the by-product of the compositional activity, we can see a one-way connection between S_0 and S_1. A graphical representation of these three operations highlights the mismatch between S_0 and S_1 (Fig. 1.8). We can gather these outcomes under the heading of *fixed use of material resources*, meaning that the dynamic of the material dimension of the creative product is completely independent from the eco-niche.

In the realm of the studio, "the composerly hand" (as Morton Feldman would put it) is too present to allow for materially grounded dynamics to shape the piece. The

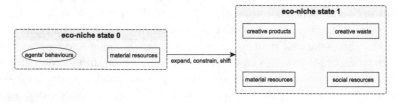

Fig. 1.8 Fixed use of material resources in sonic ecologies

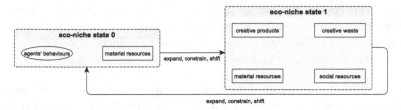

Fig. 1.9 Adaptive use of material resources in sonic ecologies

mismatch between the eco-niche and the creative procedures seems to sever the link between the composer's experience of the environment and the listener's experience. For many years, Barry Truax and other soundscape composers have been working on techniques to reconnect these two separate realities [51]. The *Urban Corridor* and *Net_Dérive* maybe signalling towards an alternative path.

The *Urban Corridor* creates a setting that enables emergent properties defined by the participants' behaviours. And *Net_Dérive* applies this approach to distributed eco-niches. From very different aesthetic perspectives—*Net_Dérive* being closer to the musique concrète tradition [47]—both pieces use accumulation as an organising strategy [32]. The material dimension of each work is moulded by the interaction among agents and objects. Event densities and timbre profiles depend both on community behaviour and on the compositional system's response to its input. Thus, the material dimension of each work is driven by the participants' behaviours. Given that these behaviours impact the sonic outcome, what we see here is a process of mutual adaptation: the eco-niche determines the material resources of the piece, but at the same time, the piece also shapes the eco-niche. These processes cannot be separated. Using strict terminology, their dynamic could be described as an *adaptive sonic ecology*: a habitat where agents and objects interact producing a creative sonic by-product that depends on local behavioural resources (Fig. 1.9).

1.4.1 Ecologically Grounded Practice and Ubimus Experimental Work

From an ecologically grounded perspective, two strategies were identified in the five musical projects discussed in this section: fixed and adaptive use of material resources. These strategies emerge from the relationships established between the local settings—more broadly defined as the ecological niche—and the decision-making processes that impacts the choice of material resources. The relationships among the material resources used during creative activities and the ecological niche have been modelled through three operations: *constrain*, *expand* and *shift*. The reviewed projects featured two approaches to the use of behavioural resources. Compositional strategies that apply a strict separation between the decision-making process and the ecological niche do not depend on the local behavioural resources. Therefore, the creative products are independent from the ecological niche. These sonic ecologies are classified as *fixed*. When the products take shape through processes that depend on the local resources, the sonic ecologies are labelled *adaptive*.

Joining the experience gained through ecologically grounded creative practice with the theoretical perspectives previously discussed provides several pointers to observables to be targeted in Ubiquitous Music experimental work. First, rather than focus on the creative outcome, experiments can target the increase of creative potentials. Potentials encompass both personal and material resources. These resources are embedded in specific social contexts. Therefore, creative activities need to be placed within ecologically valid environments in order to give access to socially meaningful results. Second, *time* becomes a key variable in assessing performance-based creativity. The flow of behavioural and material resources can be observed and manipulated to gain detailed insights on the dynamics of the creative processes. Depending on the temporal frame of the observations, experiments can target sonic events [25], musical short-term activity (Chap. 4; [33]) and longitudinal design studies [36]. Third, the unintended by-products of creative activity become as important as the intentional products. The impact of the creative processes can be assessed by observing the resources used, discarded and created during the activity as a function of time. The outcomes can be compared to the potentials assessed at the beginning of the creative cycle. If long observational windows are used, information on the sustainability of the creative approach may be retrieved (Table 1.4).

This section addressed one of the limitations observed in music-oriented models of creativity, lack of material grounding. Sonic ecologies—habitats where agents and objects interact producing creative sonic by-products—were proposed as alternatives to the disembodied contextual dimension featured in current domain-specific models. Local behavioural resources were identified as one of the important factors that characterise adaptive sonic ecologies. The flow of material and behavioural resources was proposed as the focus of performance-based creativity, highlighting time as key resource to be considered in Ubiquitous Music experiments. The comparison between the six standard creativity factors and the eco-cognitive constructs pointed to two variables that may be useful to assess the sustainability of the creative activity: (1) the *creative waste*, comprising the unintended by-products that do not serve to boost the creative potential, and (2) the *ecological footprint*, the impact of the creative action on the available resources for future activities assessed through the yield of creative waste and creative products.

The last section of this text will summarise the remaining obstacles to tackle everyday creative phenomena in the context of Ubiquitous Music research. The issues highlighted set the stage for the methodological discussion developed in the next chapter. And they point to pressing demands in the development of support infrastructure that will be partially addressed in the last three chapters of this book.

Table 1.4 Ecologically grounded creative practice: phases, constructs, observables and the 6Ps factors

Creative activity phases		Constructs	Observables	6Ps factors
Potential	1	Cognitive resources	Age, sex, schooling, musical training, technological experience	*Person*
	2	Material and social resources, affordances	Location, type and quantity of material and social resources	*Press, place*
Performance	3	Resources	Flow time, type and quantity of material, cognitive and behavioural resources	*Process (performance)*
	4	Social interaction, affordances	Time, type and quantity of behavioural resources	*Persuasion*
Outcome	5	By-products	Type and quantity of material, cognitive and behavioural resources	*Product*
			Creative waste	...
	6	Ecological niche profile	Location, type and quantity of material, cognitive and behavioural resources	*Process (outcome)*
			Ecological footprint	...
			Adaptation, new potentials and affordances	*Potential (outcome)*

1.5 Targeting the Needs of Little-c Musical Creativity

Since the late 1990s, pro-c musical practices have incorporated resources that were absent from purely instrumental music making and from studio-centred creative practice. Ubiquitous Music making has expanded this trend by embracing little-c musical phenomena that were excluded from professional musical practices. There are at least four aspects of this move towards everyday creativity that demand a stronger theoretical framework.

1. *Change of focus from creative products to processes*: Creative music making has incorporated the creative experience as a target of compositional practice. Synchronous musical phenomena cannot be reduced to a single instance or to an isolated creative product. Thus, new analytical and epistemological approaches are necessary to deal with music making as a creative experience [30, 40].
2. *Increased reliance on information technology support*: From the early 1990s, music making through technological means has become the rule, rather than the exception. Three factors contributed to the increased reliance of musical activities on technological infrastructure: (a) the wide availability of consumer

hardware (first desktop computers and later portable devices with basic sound support); (b) open source software development and unrestricted distribution of resources through the World Wide Web; and (c) knowledge networks—such as communities of practice and communities of interest—that provide both general and domain-specific support [41].

3. *Increased importance of local resources in creative activities*: During the late 1960s and early 1970s, this tendency was pioneered by site-specific art [16] and by Soundscape Composition [51]. Until recently, place-related practices were not approached from a creativity-oriented perspective [27]. Although unified place-practice creative methods are still rare, emerging paradigms—such as eco-composition [3,7,25]—are gaining strength.

4. *A shift from prescriptive models to descriptive and predictive models*: With the progressive dismissal of purely formalist approaches to music making, prescriptive compositional methods start to show their limitations. Feldman's [14] critical metaphor of "the composerly hand" summarises a general trend towards reliance on extra-musical processes (computational tools, environmental sounds, extra-musical media and audience participation) that erode the image of the isolated composer creating music just in the head. Ubiquitous Music research may provide a window to creative phenomena that lie beyond the digital instrument, the isolated sound object [47] and the isolated genius [53].

Conclusions

In this chapter, we introduced a number of important concepts in the area of creativity studies that underpin the research in UbiMus. We explored the key theoretical frameworks for the study of creativity, from the general-purpose descriptions of Runco and Glăveanu to the domain-specific principles delineated by several authors for the field of music. These were shown to inform the research in UbiMus in a very comprehensive way. Complementing this overview, we have looked at ways to enable ecologically grounded creative practices, providing some classic examples and discussing strategies for experimental work. We close the chapter by exploring ways to target the needs of everyday creative experiences.

References

1. Amabile, T.: Creativity in Context. Westview Press, Boulder (1996). http://books.google.com. br/books?id=hioVn_nl_OsC
2. Barreiro, D.L., Keller, D.: Composição com modelos sonoros: fundamentos e aplicações eletroacústicas. In: Keller, D., Budasz, R. (eds.) Criaccão Musical e Tecnologias: Teoria e Prática Interdisciplinar, pp. 97–126. ANPPO, Goiânia (2010)

3. Basanta, A.: Syntax as sign: the use of ecological models within a semiotic approach to electroacoustic composition. Organised Sound **15**(2), 125–132 (2010). doi:10.1017/S1355771810000117. http://dx.doi.org/10.1017/S1355771810000117
4. Beghetto, R.A., Kaufman, J.C.: Toward a broader conception of creativity: a case for "mini-c" creativity. Psychol. Aesthet. Creat. Arts **1**(2), 73–79 (2007)
5. Bennett, S.: The process of musical creation: interview with eight composers. J. Res. Music Educ. **24**(1), 3–13 (1976). doi:10.2307/3345061. http://jrm.sagepub.com/content/24/1/3.abstract
6. Burnard, P., Younker, B.A.: Problem-solving and creativity: insights from students? Individual composing pathways. Int. J. Music Educ. **22**, 59–76 (2004)
7. Burtner, M.: Ecoacoustic and shamanic technologies for multimedia composition and performance. Organised Sound **10**, 3–19 (2005). doi:10.1017/S1355771805000622. http://journals.cambridge.org/article_S1355771805000622
8. Cádiz, R.F.: Musical creation in the post-digital era (creación musical en la era postdigital). Aisthesis **52**, 449–475 (2012). doi:10.4067/S0718-71812012000200023. http://www.scielo.cl/pdf/aisthesis/n52/art23.pdf
9. Capasso, A., Keller, D., Tinajero, P.: The urban corridor (corredor urbano). CU Art Galleries, Boulder (2000). http://www.patriciatinajerostudio.com/collaborations/urban-corridor
10. Chen, C.W.: The creative process of computer-assisted composition and multimedia composition visual images and music. Ph.D. thesis, Royal Melbourne Institute of Technology (2006). http://researchbank.rmit.edu.au/eserv/rmit:6301/Chen.pdf
11. Collins, D.: A synthesis process model of creative thinking in music composition. Psychol. Music **33**(2), 193–216 (2005). doi:10.1177/0305735605050651. http://pom.sagepub.com/content/33/2/193.abstract
12. Di Scipio, A.: Émergence du son, son d'emergence: essai d'épistémologie expérimentale par un compositeur. Intellectica **48–49**, 221–249 (2008)
13. Dingwall, C.: Rational and intuitive approaches to music composition: the impact of individual differences in thinking/learning styles on compositional processes. Bachelor of music. University of Sydney, Sydney (2008). http://ses.library.usyd.edu.au/bitstream/2123/3991/1/Dingwall%202008.pdf
14. Feldman, M., Friedman, B.: Give my Regards to Eighth Street: Collected Writings of Morton Feldman. Exact Change, Cambridge (2000). http://books.google.ie/books?id=hfgHAQAAMAAJ
15. Ferraz, S., Keller, D.: Preliminary proposal of the mdf model of collective creation (mdf: Proposta preliminar do modelo dentro-fora de criação coletiva). Cadernos de Informática **8**(2), 57–67 (2012)
16. Friedman, K., Smith, O.F., Sawchyn, L.: The Fluxus Performance Workbook. Performance Research, vol. 7 (No. 3: 'On Fluxus'). Swinburne University of Technology, Melbourne. Performance Research e-Publications (2002). http://hdl.handle.net/1959.3/58863
17. Gabora, L., Kaufman, S.: Evolutionary perspectives on creativity. In: Kaufman, J., Sternberg, R. (eds.) The Cambridge Handbook of Creativity, pp. 279–300. Cambridge University Press, Cambridge (2010). http://arxiv.org/ftp/arxiv/papers/1106/1106.3386.pdf
18. Gibson, J.J.: The Ecological Approach to Visual Perception. Houghton Mifflin, Boston (1979)
19. Glăveanu, V.P.: Rewriting the language of creativity: the five a's framework. Rev. Gen. Psychol. **17**(1), 69–81 (2013). doi:10.1037/a0029528
20. Guilford, J.P.: The Nature of Human Intelligence. McGraw-Hill Series in Psychology. McGraw-Hill, Madison (1967). http://books.google.com.br/books?id=T-ZJAAAAMAAJ
21. Haslam, S.A., Adarves-Yorno, I., Postmes, T., Jans, L.: The collective origins of valued originality: a social identity approach to creativity. Pers. Soc. Psychol. Rev. **17**(4), 384–401 (2013). doi:10.1177/1088868313498001. http://psr.sagepub.com/content/17/4/384
22. Hickey, M.: Creative thinking in the context of music composition. In: Hickey, M. (ed.) Why and How to Teach Music Composition: A New Horizon for Music Education, pp. 31–53. The National Association for Music Education, Reston (2003)
23. Kaufman, J.C., Beghetto, R.A.: Beyond big and little: the four c model of creativity. Rev. Gen. Psychol. **13**(1), 1–12 (2009)

24. Keller, D.: touch'n'go: ecological models in composition. Master's thesis, Simon Fraser University, Burnaby (1999). http://www.sfu.ca/sonic-studio/srs/EcoModelsComposition/Title.html
25. Keller, D.: Compositional processes from an ecological perspective. Leonardo Music J. **10**, 55–60 (2000). doi:10.1162/096112100570459. http://muse.jhu.edu/journals/leonardo_music_journal/v010/10.1keller.pdf
26. Keller, D.: Metrophonie. In: CD Music from Stanford 541, vol. 1. Innova, St. Paul (2005). Http://www.innova.mu/sites/www.innova.mu/files/liner-notes/635.htm
27. Keller, D.: Sonic ecologies. In: Brown, A.R. (ed.) Sound Musicianship: Understanding the Crafts of Music, pp. 213–227. Cambridge Scholars Publishing, Newcastle upon Tyne (2012). http://www.c-s-p.org/flyers/978-1-4438-3912-9-sample.pdf
28. Keller, D., Berger, J.: Everyday sounds: synthesis parameters and perceptual correlates. In: Proceedings of the Brazilian Symposium on Computer Music (SBCM 2001). SBC, Fortaleza (2001). http://ccrma.stanford.edu/~dkeller/pdf/EverydaySounds.pdf
29. Keller, D., Capasso, A.: New concepts and techniques in eco-composition. Organised Sound **11**(1), 55–62 (2006). doi:10.1017/S1355771806000082. http://dx.doi.org/10.1017/S1355771806000082
30. Keller, D., Ferneyhough, B.: Analysis by modeling: Xenakis's st/10-1 080262. J. New Music Res. **33**(2), 161–171 (2004). doi:10.1080/0929821042000310630. http://www.tandfonline.com/doi/abs/10.1080/0929821042000310630
31. Keller, D., Truax, B.: Ecologically based granular synthesis. In: Proceedings of the International Computer Music Conference, pp. 117–120. MPublishing, University of Michigan Library, Ann Arbor (1998). http://quod.lib.umich.edu/cgi/p/pod/dod-idx/ecologically-based-granular-synthesis.pdf
32. Keller, D., Capasso, A., Wilson, S.R.: Urban corridor: accumulation and interaction as form-bearing processes. In: Proceedings of the International Computer Music Conference (ICMC 2002), pp. 295–298. MPublishing, University of Michigan Library, Ann Arbor (2002). http://quod.lib.umich.edu/cgi/p/pod/dod-idx/urban-corridor-accumulation-and-interaction-as-form-bearing.pdf?c=icmc;idno=bbp2372.2002.061
33. Keller, D., Barreiro, D.L., Queiroz, M., Pimenta, M.S.: Anchoring in ubiquitous musical activities. In: Proceedings of the International Computer Music Conference, pp. 319–326. MPublishing, University of Michigan Library, Ann Arbor (2010). http://hdl.handle.net/2027/spo.bbp2372.2010.064
34. Keller, D., Lima, M.H., Pimenta, M.S., Queiroz, M.: Assessing musical creativity: material, procedural and contextual dimensions. In: Proceedings of the National Association of Music Research and Post-Graduation Congress—ANPPOM (Anais do Congresso da Associação Nacional de Pesquisa e Pós-Graduação em Música—ANPPOM), pp. 708–714. National Association of Music Research and Post-Graduation (ANPPOM), ANPPOM, Uberlândia (2011). http://www.anppom.com.br/anais.php
35. Keller, D., Ferreira da Silva, E., Pinheiro da Silva, F., Lima, M.H., Pimenta, M.S., Lazzarini, V.: Everyday musical creativity: an exploratory study with vocal percussion (criatividade musical cotidiana: Um estudo exploratório com sons vocais percussivos). In: Proceedings of the National Association of Music Research and Post-Graduation Congress—ANPPOM (Anais do Congresso da Associação Nacional de Pesquisa e Pós-Graduação em Música—ANPPOM). ANPPOM, Natal (2013). http://anppom.com.br/congressos/index.php/ANPPOM2013/Escritos2013/paper/view/2098/420
36. Keller, D., Otero, N., Pimenta, M.S., Lima, M.H., Johann, M., Costalonga, L., Lazzarini, V.: Relational properties in interaction aesthetics: the ubiquitous music turn. In: Proceedings of the Electronic Visualisation and the Arts Conference (EVA-London 2014). Computer Arts Society Specialist Group, London (2014)
37. Kozbelt, A., Beghetto, R.A., Runco, M.A.: Theories of creativity. In: Kaufman, J.C., Sternberg, R.J. (eds.) The Cambridge Handbook of Creativity, Cambridge Handbooks in Psychology. Cambridge University Press, Cambridge (2010). http://pages.uoregon.edu/beghetto/CreativityTheories(Kozbelt,Beghetto%26Runco).pdf

38. Lazzarini, V., Yi, S., Timoney, J., Keller, D., Pimenta, M.S.: The mobile csound platform. In: Proceedings of the International Computer Music Conference, pp. 163–167. ICMA, Ann Arbor, MPublishing, University of Michigan Library, Ljubljana (2012). http://quod.lib.umich.edu/cgi/p/pod/dod-idx/mobile-csound-platform.pdf
39. Lima, M.H., Keller, D., Pimenta, M.S., Lazzarini, V., Miletto, E.M.: Creativity-centred design for ubiquitous musical activities: two case studies. J. Music Technol. Educ. 5(2), 195–222 (2012). doi:10.1386/jmte.5.2.195_1. http://www.ingentaconnect.com/content/intellect/jmte/2012/00000005/00000002/art00008
40. Marsden, A.: What was the question? Music analysis and the computer. In: Crawford, T., Gibson, L. (eds.) Modern Methods for Musicology, pp. 137–147. Ashgate, Farnham (2009). http://www.lancs.ac.uk/staff/marsdena/publications/MarsdenForEPrints.pdf
41. Pimenta, M.S., Miletto, E.M., Keller, D., Flores, L.V.: Technological support for online communities focusing on music creation: adopting collaboration, flexibility and multi-culturality from Brazilian creativity styles. In: Azab, N.A. (ed.) Cases on Web 2.0 in Developing Countries: Studies on Implementation, Application and Use, Chap. 11. IGI Global Press, Vancouver (2012). doi:10.4018/978-1-4666-2515-0.ch011. http://www.igi-global.com/chapter/technological-support-online-communities-focusing/73062
42. Pinheiro da Silva, F., Keller, D., Ferreira da Silva, E., Pimenta, M.S., Lazzarini, V.: Everyday musical creativity: exploratory study of ubiquitous musical activities (criatividade musical cotidiana: estudo exploratório de atividades musicais ubíquas). Música Hodie 13, 64–79 (2013)
43. Reich, S.: It's gonna rain [electroacoustic]. Compact Disc Steve Reich: Early Works [Nonesuch 9 79169-2] (1965/1987)
44. Rhodes, M.: An analysis of creativity. Phi Delta Kappan 42, 305–311 (1961). http://www.jstor.org/stable/20342603
45. Richards, R., Kinney, D., Benet, M., Merzel, A.: Assessing everyday creativity: characteristics of the lifetime creativity scales and validation with three large samples. J. Pers. Soc. Psychol. 54, 476–485 (1988)
46. Runco, M.A.: A hierarchical framework for the study of creativity. New Horizons Educ. 55(3), 1–9 (2007). http://files.eric.ed.gov/fulltext/EJ832891.pdf
47. Schaeffer, P.: Traité des objets musicaux: Essai interdisciplines. Éditions du Seuil, Paris (1977)
48. Shah, J.J., Smith, S.M., Vargas-Hernández, N.: Metrics for measuring ideation effectiveness. Des. Stud. 24(2), 111–134 (2003). doi:10.1016/S0142-694X(02)00034-0. http://people.tamu.edu/~stevesmith/SmithCreativity/ShahVargas-HernandezSmith2002.pdf
49. Simonton, D.K.: History, chemistry, psychology, and genius: an intellectual autobiography of historiometry. In: Runco, M., Albert, R. (eds.) Theories of Creativity, pp. 92–115. Sage Publications, Newbury Park (1990)
50. Tanaka, A., Gemeinboeck, P.: A framework for spatial interaction in locative media. In: New Interfaces for Musical Expression (NIME), pp. 26–30 (2006)
51. Truax, B.: Genres and techniques of soundscape composition as developed at simon fraser university. Organised Sound 7(1), 5–14 (2002). doi:10.1017/S1355771802001024. http://dx.doi.org/10.1017/S1355771802001024
52. Webster, P.R.: Asking music students to reflect on their creative work: encouraging the revision process. In: Yip, L.C.R., Leung, C.C., Lau, W.T. (eds.) Curriculum Innovation in Music (4th Asia-Pacific Symposium on Music Education Research), pp. 16–27. The Hong Kong Institute of Education, Hong Kong (2003). doi:10.1080/1461380032000126337. http://dx.doi.org/10.1080/1461380032000126337
53. Weisberg, R.W.: Creativity: Beyond the Myth of Genius. Books in Psychology. W. H. Freeman, New York (1993). http://books.google.com.br/books?id=_VNwQgAACAAJ
54. Westerkamp, H.: Kits beach soundwalk [for spoken voice and two-channel tape]. Compact Disc Transformations (1989/1996). http://www.sfu.ca/~westerka/program_notes/kits.html
55. Yamamoto, Y., Nakakoji, K.: Interaction design of tools for fostering creativity in the early stages of information design. Int. J. Human-Comput. Stud. 63(4–5), 513–535 (2005). doi:10.1016/j.ijhcs.2005.04.023. http://www.sciencedirect.com/science/article/pii/S1071581905000480

Chapter 2
Methods in Creativity-Centred Design for Ubiquitous Musical Activities

Marcelo S. Pimenta, Damián Keller, Luciano V. Flores,
Maria Helena de Lima, and Victor Lazzarini

Abstract In this chapter we describe a set of creativity-centred design methods including strategies for interaction, signal processing, planning, prototyping and creativity assessment. Social, material and procedural requirements were gathered through a ten-subject planning design study. Based on these results, an interaction metaphor—time tagging—was developed to deal with a musical activity in ubiquitous contexts: localised audio mixing. We implemented a series of prototypes—the first generation of mixDroid—for mixing using Android-based mobile devices. An exploratory field study with mixDroid was conducted inside the studio and in the locations where the sound samples were recorded. Activities happening outside the studio resulted in higher creativity scores on two dimensions—explorability and productivity. We discuss the preliminary implications of these findings for future experiments targeting aspects of exploratory creativity in everyday settings.

2.1 Introduction

Our research group has centred its efforts on the development of computing technology to support both materially grounded artistic practices and novice-oriented computer-based musical activities. This initiative has followed an interdisciplinary approach, involving a multidisciplinary team of experts in computer music, eco-composition, music education, human–computer interaction (HCI), ubiquitous computing (ubicomp) and computer supported cooperative work (CSCW), pointing toward a new research field: Ubiquitous Music. We have proposed the adoption of

M.S. Pimenta (✉) • L.V. Flores • M.H. de Lima
Federal University of Rio Grande do Sul, 91501-970 Porto Alegre, RS, Brazil
e-mail: mpimenta@inf.ufrgs.br; lvflores@inf.ufrgs.br; helena.lima@ufrgs.br

D. Keller
Amazon Center for Music Research - NAP, Federal University of Acre, Rio Branco, Brazil
e-mail: dkeller@ccrma.stanford.edu

V. Lazzarini
Music Department, Maynooth University, Ireland
e-mail: vlazzarini@nuim.ie

© Springer International Publishing Switzerland 2014
D. Keller et al. (eds.), *Ubiquitous Music*, Computational Music Science,
DOI 10.1007/978-3-319-11152-0_2

the term Ubiquitous Music (or simply UbiMus) to promote practices that empower participants of musical experiences through socially oriented, creativity-enhancing tools [25]. To achieve this goal, our group has been engaged in a multidisciplinary effort to investigate the creative potential of converging forms of social interaction, mobile and distributed technologies and innovative music-making practices.

Our goals have pushed us toward participatory research methodologies, focusing on human aspects within the field of Information Technology Creative Practices [35]. The various methods that we have applied to the development and assessment of technological, educational and artistic outcomes can be grouped under the common denominator of *creativity-centred design techniques*. Recent results point to a particularly promising area of application of creativity-centred design, targeting forms of creativity accessible to non-specialised users performing Ubiquitous Musical activities outside of domain-specific venues.

This chapter highlights the close relationships between Ubiquitous Music research and *everyday creativity*. First, we discuss recent advances in general creativity research, defining the domain of application of everyday creativity (Chap. 1; [42, 43]) in the context of Ubiquitous Music research. Ubiquitous musical activities open opportunities for musical creation by musicians and untrained participants outside studio settings. Given the specific demands of the experimental work in these harsh contexts, strategies that enable data collection without disrupting the creative experience become a requirement. Our previous work pointed toward three possible solutions for this methodological conundrum: (1) avoiding early commitment to specific tools [14, 26], (2) supporting iterative development through rapid prototyping [27] and (3) fostering collaboration by building communities of practice [32, 34].

Creativity-centred design of Ubiquitous Musical systems involves at least four developmental stages: (1) *defining strategies for design*, (2) *planning*, (3) *prototyping* and (4) *assessment*. The definition of strategies for design is concerned basically with selection and adoption of interaction metaphors and patterns. Our group has been investigating artistic and educational applications of methods based on HCI and ubiquitous computing techniques (see Chaps. 3 and 4 for related works). *Metaphors* for interaction provide abstractions that encapsulate solutions applicable to a variety of activities without enforcing unnecessary technical restrictions [37,38]. Thus, interaction metaphors can be built from general ergonomic principles that take into account both human and technological aspects of the activity. On a similar vein, recurring technological solutions can be grouped as *interaction patterns* [14]. These patterns are particularly useful when developers face the task of finding suitable techniques to deal with specific interface implementation issues. Furthermore, technologically based musical environments also demand tailoring support for sound rendering. *Signal-processing* techniques have to be chosen according to the characteristics of the task, the computational resources of the infrastructure and the profile of the target users. Ubiquitous Musical activities may involve mobility, connectivity and coordination among heterogeneous devices with scarce computational resources. Thus, carefully chosen software design strategies are a prerequisite to tackle signal processing in ubiquitous contexts (Chapters 6 and 7 ; [31]).

Ubiquitous Music planning studies involve early assessment of target population expectations and identification of opportunities for creativity support. Through a UbiMus *planning study*, Lima and coauthors [32] found sharply differing expectations on technological usage by musicians and musically naive subjects in educational contexts. Based on these results, they proposed a simple rule of thumb: users like what comes closer to reenacting their *previous musical experience*. Nontechnical approaches, such as those proposed by traditional soundscape activities [44], may not be suited for introducing nonmusicians to sonic composition. Naive subjects may respond better to technologically oriented approaches, such as those employed in eco-composition [21, 32]. If the rule of thumb previously stated holds true, musically untrained participants would welcome easiness of use and naturality, while musicians would tend to prefer interfaces that reproduce behaviours based on *acoustic-instrumental metaphors* and common-practice *music notation*. Therefore, design of creatively oriented technologies would need to fulfil different demands depending on the intended user base.

Technological support for pervasive musical activities increases the difficulty of the design task on two fronts. Ubimus systems may enhance the users' creative potential by providing access to material and social resources. But a wider accessibility to resources could bring up unintended issues that limit the systems' adoptability to a small user base. Thus, the challenge of UbiMus design to provide intuitive tools for complex creative tasks does not guarantee wide accessibility. Custom-made, special purpose hardware interfaces may fill the requirements of transparency and naturalness reducing the cognitive load of complex tasks [13]. But these systems are difficult to distribute and maintain. So, the user base is narrowed by the increased costs of the hardware. Two techniques that we have been applying may provide viable solutions to the problem of sustainability without negative impacts on usability: hardware repurposing [19] and rapid prototyping. We have reutilised consumer-class mobile devices, such as cellphones and tablets, as creative musical tools [14, 24, 31]. Within an iterative approach to design, involving creative musical activities and usability assessments, we have developed rapid prototyping techniques tailored for ubiquitous contexts. Since our research targets interaction and signal processing, flaws that arise from coordination among these two processes can be identified early within the design cycle.

Prototyping encompasses mainly two areas of expertise: interaction and signal processing, as mentioned above. Technological support for pervasive musical activities increases the difficulty of the design task on two fronts. Ubimus systems may enhance the users' creative potential by providing access to material and social resources. But a wider accessibility to resources could introduce unintended complexities that limit the systems' adoptability to a small user base. Thus, one challenge of UbiMus design is to provide *intuitive tools for complex creative tasks*. Nevertheless, attaining this objective does not guarantee wide accessibility. Custom-made, special purpose hardware interfaces—such as those proposed by research in tangible user interface design—may fill the requirements of transparency and naturality reducing the cognitive load of complex tasks [13]. In this case, the catch lies in the financial toll. Special-purpose systems are difficult to distribute and maintain. Consequently, the user base is narrowed by the increased costs of the hardware. A set of strategies adopted by our group may provide viable solutions to the problem of sustainability without a negative impact on usability. Two techniques that we have been applying are hardware *repurposing* [19] and *rapid prototyping*. We have reutilised consumer-class mobile devices—such as cellular telephones and portable tablets—as creative musical tools. Given the lack of standard support for audio and musical data formats, initial development for mobile platforms was feasible but complex and unintuitive [14, 24, 46]. Recent advances have paved the way to wider usage within the computer music community [4, 11, 31]. Within an *iterative approach to design*—involving creative musical activities and usability assessments—we have developed rapid prototyping techniques tailored for ubiquitous contexts. Since our research targets interaction and signal processing, flaws that arise from coordination among these two processes can be identified early within the design cycle. Furthermore, on-site usage in full-blown musical activities uncovers opportunities for creative exploration of the software *and* of the environment.

The fourth stage of the design cycle involves the empirical assessment of the design proposals. Although creativity assessment is an expanding area of research within psychology [1, 36, 40], assessment of creative outcomes is still a taboo topic among music practitioners. From a universalist perspective, creativity assessment would be equivalent to measurement of musical value. Standards are defined by the adopted compositional technique. Given an adopted metric, deviations from the standard are seen as spurious, less valuable manifestations. This approach makes two assumptions. First, the objective of musical activity is to obtain a product that can be labelled as an expression of eminent creativity. Second, the value of the musical product lies in its material constituents (the sounds or their symbolic representation, i.e. scores or recordings), not on the extramusical aspects of music making. A second limitation of creativity assessments is the overrated reliance on expert judgement within the musical domain. When asked to evaluate musical products, as it is done using Amabile's [1] Consensual Assessment Technique, experts apply their views on creativity. These views are the result of several years of musical training and experience with eminent forms of creativity. Therefore, unless they re-educate themselves to disregard their background, expert evaluators will tend to assess musical products by eminent-creativity standards. This bias renders their evaluations less useful to everyday creativity manifestations. To avoid these pitfalls, we rely on a mix of assessment techniques, or a "triangulation", within the behavioural research literature, engaging a small number of expert and untrained subjects in different musical activities under a variety of environmental conditions. This approach does not make assumptions regarding the compositional techniques and assigns the same weight to musicians' and lay people's feedback. Given a support infrastructure under study, what we look for are the relationships among user profiles, types of activity and environmental conditions.

The contribution of this chapter to the field of new media technologies is threefold. On the one hand, we define a clear target for Ubiquitous Music research endeavours: to *enhance everyday creativity*. On the other, we bring together a scattered body of methodological knowledge to coalesce into a coherent set of procedures: creativity-centred *design for musical activities*. And as an example of the proposed approach, we describe the exploratory creativity assessment of a Ubiquitous Music interaction metaphor: time tagging.

2.2 Ubiquitous Music and Everyday Creativity

This section focuses on general creativity theories that can be applied to the study of Ubiquitous Musical practices (see Chap. 1 for an in-depth discussion of creativity theories in the context of UbiMus research). The concepts proposed are not domain specific; therefore—although our target application is creative musical activity—the perspectives put forth are also applicable to performance art practices, education and computing. Particular care is taken to avoid technical assumptions that would narrow the applicability of the concepts proposed. Our objective is to situate everyday creativity within the context of general creativity research.

According to the discussion in Chap. 1 (Sect. 1.2), creativity magnitudes encompass four levels: Big-C or eminent creativity, pro-c or professional creativity and two forms of everyday creativity—little-c and mini-c—which correspond to mundane creative experiences that yield creative products and personal experiences of creativity without material products. Although some forms of Ubiquitous Music making may attain pro-c level, the greatest potential of novice-oriented technological support lies in everyday manifestations of creativity. Furthermore, musical activities that take place outside the regular specialised venues—such as eco-composing—until recently demanded special purpose, usually expensive portable infrastructure (see [22]). By providing accessibility to outside locations, UbiMus practice highlights the importance of the place c-factor within everyday creative endeavours.

Our research on Ubiquitous Music practices has uncovered two areas of intersection with little-c studies. Ubiquitous Music engages participants in creative activities that do not demand professional-level musical training or long-term learning of technological tools. Indeed, in recent years, we have investigated the support for creativity of novice (layman) people. Like Weinberg [48], we are interested in providing access to meaningful and engaging musical experiences to both novices in music and experienced musicians (see Chap. 4 for a music education example). Novices generally do not act or think like musicians. Novices usually do not have access to musical instruments or to the knowledge required to play them. Novices do not know how to apply compositional techniques, and they are discouraged by the complexities of common-practice notation. In fact, the large corpus of formal knowledge associated to acoustic-instrumental musical composition, in general, acts as a barrier to creative activities by novices. Clearly, designing for music creation by novices and for instrumental music composition activities is not the same [33].

2.3 Mobility and Connectivity in Ubiquitous Music

By incorporating previously unreachable resources, ubiquitous computing systems provide two features that in conjunction expand the potential of creative activities: mobility and connectivity [49]. Mobile devices are carried by users to multiple places and are routinely incorporated into everyday activities. Whether connected to wireless networks or used as autonomous devices, portable telephones, tablets and personal digital assistants carry enough computer power to allow for complex musical activities lasting from a few minutes to several hours. Within urban settings, portable devices can make use of remote resources, thus expanding their context-awareness capabilities [9]. If activities are restricted to locations where network access is available, fully connected ubiquitous devices can act as clients of remote systems. In this scenario—usually referred as cloud computing—a key limitation is the bandwidth and quality of the data-exchange service which may introduce jitter and delay. Activities that demand permanent synchronous access to remote resources may be limited to locations that feature high-quality connectivity. But if the synchronisation requirements are relaxed—as it is the case in asynchronous systems—areas that provide intermittent remote access may also enhance their resources by updates based on network availability. Given these two essential features of the ubiquitous support infrastructure—mobility and connectivity—Ubiquitous Musical activities may be classified according to their demand of resources. On-site activities make use of material resources locally available (we call these *just plain objects*—JUPOS). When the activity engages more than one stakeholder, exchange of social resources also takes place. In this case, social resources are classified as *just plain agents* (or JUPAS). Sometimes, local activities will involve technological mediation to gain access to material resources. In this scenario, computationally mediated material resources are labelled ITOS or *I*nformation *T*echnology-mediated *O*bjects. But for the activities that use nonlocal resources, only material resources supported through connectivity are accessible. Therefore, these activities necessarily demand support infrastructure for ITOS. On a similar vein, activities that make use of social resources from remote locations provide access to *I*nformation *T*echnology-mediated *A*gents (or ITAS). The remaining resources used in UbiMus activities belong to the realm of the individual, so they do not require technological support to become available. These are called *personal resources* (or self). Table 2.1 summarises the relationships between resource accessibility and technological support in Ubiquitous Musical activities. This summary clarifies the impact of mobility and connectivity on the requirements for everyday creativity. Through mobility, ITOS can now be combined

Table 2.1 Resource accessibility and technological support in Ubiquitous Musical activities

Technological support	Resources	Acronyms
Mobility	Local material resources	JUPOS—just plain objects
Mobility	Local social resources	JUPAS—just plain agents
Connectivity	Remote social resources	ITAS—information technology mediated agents
(Optional)	Personal resources	Self

with locally available JUPOS within the same activity. Thus, visual, sonic and haptic cues which are location specific can be used to enhance the creative possibilities of on-site musical activities. More to the point, cues can be created to modify the environmental characteristics to suit the needs of the activity. This "re-engineered" environment becomes the interface for musical activities. In a sense, the concept of interface becomes meaningless since while the environment acts upon the agents, the agents modify the environment. This process of mutual adaptation is encapsulated by the concept of *anchoring* [20, 24]. But just like mobility is important to provide access to local resources, connectivity is required to bring remote objects and agents into the creative activity. While mobility allows agents to interact with ITOS in previously inaccessible locations, connectivity provides a bridge between local and remote resources. Agents in remote locations can contribute their social resources to the creative activity. Material resources from different locations can be combined synchronously—in the case of fully connected systems—or asynchronously—for creative activities that do not require immediate feedback. Thus, while the greatest impact of mobility is on JUPOS and JUPAS, connectivity enhances the usage of ITOS and ITAS.

The existence of technological support and the availability of resources are not enough to guarantee the incorporation of the full potential of ubiquitous computing infrastructure into creative artistic practices. Creativity support tools demand strategies that target usage in actual settings [45], and Ubiquitous Music systems are no exception. Building on top of previous work on eco-composition [23] and collaborative music making [34], our group has enlarged the toolbox of methodological strategies for creativity-centred software design. Bringing together the concepts discussed in this section, we can state that *Ubiquitous Music is a research field that deals with systems of human agents and material resources that afford musical activities through sustainable creativity support tools*. Mobility and connectivity are two aspects of the technological infrastructure that impact how resources are handled during the creative activities. These are basic but not sufficient conditions for artistic creation in pervasive contexts. In order to enable manifestations of everyday creativity in music, it is necessary to design support for creative musical activities in everyday settings.

2.4   Creativity-Centred Design

Só far, we have discussed how ubiquitous computing infrastructure impacts music making by providing or restricting access to social and material resources. We introduced a definition of Ubiquitous Music that takes human activity as a unit of study, highlighting its material grounding and pointing to creativity and sustainability as two long-term objectives of technologically based artistic research. We also situated UbiMus practice within the context of everyday creativity. Now we will discuss the implications of this conceptual framework within the context of creativity-centred design.The procedures that have emerged for supporting Ubiquitous Musical activities encompass the four interrelated stages of defining design strategies, planning, prototyping and assessment, as discussed in the introduction. Given the iterative and participatory nature of our design practice, these four stages are not necessarily successive, and each stage may be repeated several times during the development cycle (Fig. 2.1).

Our practice suggests three emergent methodological trends that may be used as general guidelines to define design strategies:

(a) avoid early domain restriction;
(b) support rapid prototyping;
(c) foster social interaction.

After the initial choice of design strategies, planning activities may be pursued in the form of exploratory studies. The objective of this design phase is to obtain a set of requirements and to gather initial feedback on user expectations. Once the minimal requirements and the overall objectives of the project have been set, simple prototypes can be built to allow for more detailed on-site observations. Prototypes do not need to be complete software solutions. This stage's objective is to gather useful information on specific aspects of the musical experience. Thus, sonic outcomes can be handled by simplified signal-processing tools [31] or by Wizard of Oz simulations [16]. Design issues of the adopted interaction approach can be studied by using software mash-ups, verbal scores, aural scores, graphic scores, storyboards, videos and animations. The focus is fast turn-over,

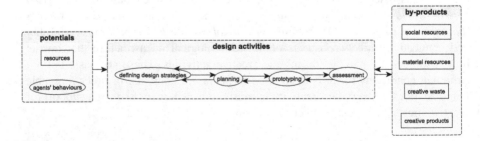

Fig. 2.1 The current status of creativity-centred design

not refined implementations. Field trials come at a late phase of the design process and preferably involve small groups of subjects doing multiple activities in various realistic contexts. The main objective is to gather data on the impact of the environmental context on the musical experience. This is tricky, and no ready-made recipes are available yet. Assessment should be as closely tied to the activity as conditions permit. Both objective data, related to the subjects' profile, activity variables, environmental variables and technological infrastructure, and subjective data, the subjects' feedback on various aspects of the experience, should be gathered. Through comparisons among various conditions, it is possible to evaluate the impact of the material and the social resources on the participants' performance. These results feed the previous design phases, pointing to updated strategies and prototype refinements.

2.5   Phase 1: Defining Strategies for Ubimus Design

2.5.1   Avoid Early Domain Restriction

Music creativity models that emphasise the material dimension provide the most direct window to experimental observation. We define the material dimension as the collection of resources available to the participants of a creative activity. In this case, we are dealing with sound sources or tools used to generate creative musical products. Thus, the connection to material resources is direct. As discussed in Chap. 1, two of the three interrelated stages suggested by Dingwall [10], the generation stage and the development stage, can easily be assessed by measuring the quantity of the material produced. The final stages involve selection, grouping and disposal of material resources; therefore, both objective and subjective assessments may be necessary. Subjective assessment of creative products can be done through Amabile's [1] Consensual Assessment Technique, CAT. Objective assessment demands measurements of the resource yield and the re-source consumption as a function of time [12]. Regarding material resources, Bennett's [2] model suggests that compositional processes start from a single germinal idea. Collins [8] also adopts this view but proposes multiple materials rather than a single idea. Contrastingly, Hickey [17], Burnard and Younker [5], Chen [7] and Dingwall [10] models suggest that exploratory activities precede the selection of materials.

The methodological difficulty resides in the task choice for creativity assessment experiments. The underlying hypothesis is, as suggested by Hickey, Burnard and Younker, Chen and Dingwall models, that both restricting and providing access to materials are part of the compositional process. Therefore, by selecting materials or tools, the experimenter is taking the place of the composer, and the resulting data cannot be used to determine whether the creative musical activity begins by exploratory actions or by a given set of materials. If the musical materials are given by the experimenter, it will not be possible to draw any conclusions regarding the initial handling of material resources. We label this problem as *early domain restriction* (Chap. 1; [26, 32]).

The most important message to be drawn from the studies discussed in this section is that the experimental settings do interfere with the creative experience. Laboratory-based studies, early choice of tools and design driven by compositional techniques, if not explicitly treated as experimental variables, may limit the applicability of the results on the material dimension.

2.5.2   Support Rapid Prototyping

A difficulty faced by the designers of musical tools is the slowness of the validation cycle. Because complete integrated systems are hard to design and test, tools usually deal with isolated aspects of musical activity. Musicians' usage of the tools may not correspond to the intended design, and integration of multiple elements may give rise to unforeseen problems. As a partial solution to these hurdles, we have suggested the adoption of rapid prototyping and the inclusion of music making within its development cycles. This integration of music making and software development is based on a broad approach to usability [3,18]. Fine-grained technical decisions are done after the usability requirements of the system have been well established through actual usage. So rapid deployment is prioritised over testing on a wide user base.

2.5.2.1   Prototyping Creative Interaction

If we see music creation as a design activity, it seems natural and straightforward to adopt a prototypical approach to the study of factors that shape this creative process. *Draft* is a term commonly applied to an unfinished musical product, but since our methodological emphasis is on the cyclical creation process rather than on the product, our group has proposed the term *musical prototype* for the creative output produced in musical activities [34]. A prototypical music creation process means that novices can draft initial musical ideas (a musical prototype) which can be tested, modified through a cyclical refinement process until a consensus is reached. This process resembles the prototyping cycles adopted in industry and in incremental software development. But instead of dealing with goals set from the start of the activity, creative prototyping involves exploration of material resources to help define partial targets and to constrain the range of creative choices. Adapting the concepts proposed by the field of distributed cognition [20], we have suggested that creative activity encompasses *epistemic activities* which help to enhance the individual's knowledge of the world and *pragmatic/enactive* activities which provide actual creative outcomes [24].

Therefore, in creativity-centred design of Ubiquitous Musical systems, we simultaneously deal with tool prototypes and musical prototypes. The former serve to test hypothesis on musical interaction processes, formulated as metaphors for interaction and interaction patterns [14, 27, 37]. The latter are used to assess the impact of the proposed technologies on the creative outcomes.

Musical prototypes are used to assess the impact of the proposed technologies on the creative outcomes. Metaphors for interaction embody methodological solutions that are not bound to technical specificities. For example, the time-tagging metaphor is applicable to mixing on stationary or on portable devices. It can be applied on sonic data or on control sequences. It could also be extended to video applications (although this usage would conflict with the objective of reducing computational-resource usage). Thus, interaction metaphors materialise general ergonomic principles to fulfil the human and the technological demands of the activity. When similar technological solutions are observed in various contexts, interaction patterns may provide a useful generalisation. Interaction patterns can be applied to the task of finding suitable techniques to deal with interface implementation issues. So far, our group's research has unveiled four musical interaction patterns: *natural interaction, event sequencing, process control* and *mixing* [15]. Each of these patterns tackles a specific interaction problem. *Natural interaction* deals with forms of musical interaction that are closely related to handling everyday objects. *Event sequencing* lets the user manipulate temporal information by freeing the musical events from their original timeline. *Process control* provides high-level abstractions of multiple parametric configurations, letting the user control complex processes by using simple actions. *Mixing* can be seen as the counterpart of event sequencing for synchronous interaction. Musical data—including control sequences and sound samples—is organised by user actions that occur in-time.

We can think of interaction metaphors and patterns as results of opposite design trends. While metaphors provide recognisable instantiations of general interaction mechanisms, patterns are reusable generalisations of specific solutions. This means that solutions encountered by inductive or bottom-up processes (patterns) could eventually match solutions reached top-down, through deduction of general principles (metaphors). These specific cases are the strongest candidates for useful applications in multiple design contexts. Thus, a possible contribution of Ubiquitous Music results to the area of design is to help identify these cases.

2.5.2.2   Prototyping Signal Processing

Interaction is a key aspect of effective support for creative music making. But giv-en their reliance on computer-based sound, Ubiquitous Musical systems also need to address design issues related to sound synthesis and processing techniques. Lazzarini and coauthors [31] report the development of the Mobile Csound Platform (MCP) to bring the Csound language to popular mobile device operating systems. Work was done to build an idiomatic, object-oriented API for both iOS and Android operating systems, implemented using their native languages (Objective-C and Java, respectively). Work was also done to enable Csound-based applications to be deployed over the internet via Java Web Start (JAWS) (see Chaps. 6 and 7 for related technologies). By porting Csound to these platforms, Csound as a whole moved from embracing usage on the desktop to becoming pervasively available.

The usage of the Csound language as well as its exposure to users has changed over the years. In the beginning, users were required to understand the specifics of Csound syntax and coding to deal with sound synthesis. Today, applications are developed that expose varying degrees of Csound coding, requirements ranging from full knowledge of Csound to none at all. Applications such as those created for the XO platform highlight a task-focused interface, leveraging Csound for its audio capabilities. Other applications—such as Cecilia—provide users with a task-focused interface, but the capability to extend the system is available only to those who know Csound coding. Thus the Csound language has grown from a means to render musical data to becoming a general-purpose language for audio engine programming and music making.

Csound was deployed as the sound engine for one of the pioneer portable systems, the XO-based computer used in the One Laptop per Child (OLPC) project [30]. The possibilities allowed by a re-engineered version of Csound were partially exploited in this system. Its development sparked several ideas for its ubiquitous usage, which is now steadily coming to fruition with a number of parallel projects, collectively named the MCP. Through a flexible approach to audio synthesis and processing, concurrent design of sound synthesis and interaction support becomes possible. Prototypes can be created on a stationary device, tested through web deployment and eventually compiled for mobile platforms. Thus, issues arising on devices with scarce computational resources can be detected and considered within a single design cycle. In particular, limitations on synchronous sound rendering may impact the quality of the musical interaction experience. This unified and iterative procedure provides a much more detailed feedback on the coordination requirements among interaction and synthesis processes. The emerging methods that enable these advances incorporate the philosophy of integrated concepts and tools development. Thus, they can be labelled *Ubiquitous Music ecosystems* [31].

2.5.3 Foster Social Interaction

One of the aspects of everyday creativity that sets it apart from Big-c approaches to music making is the distributed nature of its resources. Creativity in everyday settings demands the usage of local material resources, social interaction with unforeseen participants and quick adaptation to volatile conditions. These demands are nothing similar to the carefully isolated musical venues and the fixed roles for individuals in isolated compartments assigned to creative tasks. While Big-c relies on the composer, little-c demands the engagement of multiple (sometimes anonymous) actors for creative action. This is one of the reasons why Ubiquitous Music research is so closely related to everyday creativity investigations.

Community-based methods are at the centre of Ubiquitous Music practice. The concept of community is based on the locality of human life and social interaction. People that form a community share a space and possibly common goals. Belonging to the same community implies influencing each other, either directly (e.g. via direct communication) or indirectly (e.g. via providing and retrieving public information). Thus, supporting virtual communities is strongly related to relationship management. While communicating with each other or providing and retrieving shared information, people establish relationships with other people. This social network forms the basis for shared activities and encourages rich exchanges of material and social resources that may provide support for learning within the community.

The free access to know-how and the fast circulation of resources within social groups with common objectives foster the emergence of a phenomenon quite relevant to Ubiquitous Music research: the *communities of practice* [29]. A community of practice is a simple social system that arises out of learning and exchange processes. A key aspect of this type of community is that it unfolds through practice, not prescription [50] (p. 192), so it can be seen as an extension of the dialogical perspective that has emerged within the educational field [32]. For example, open-source communities that are agile and flexible foster engagement, imagination and alignment. We believe these characteristics may provide a fertile ground for creativity-centred design.

Summing up, social interaction happens at three levels of the creativity-centred Ubiquitous Music design cycle:

(a) as a resource for effective musical experiences;
(b) as a tool for design assessment and critical evaluation;
(c) as a factor for growth and consolidation of a community of practice engaged in Ubiquitous Music research.

The lower level, studied within the context of musical creative activities, is materialised in the musical prototyping process, fostering social exchange among music practitioners. The middle level, involving design activities that serve to adjust the objectives and the research methods, is an arena of negotiation among artistic, computational and educational perspectives. The higher level, the community of practice, encompasses both novice practitioners and designers, providing circulation of material and social resources that foster the community's growth.

2.6   Phase 2: Planning

The UbiMus planning protocol's objective is to obtain a set of social, procedural and material requirements to be applied in the design of creativity-centred systems. Social aspects of creative activities are related to the interactions among agents and to the factors that influence the dynamic of the creative process. Technological

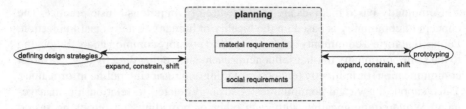

Fig. 2.2 Ubimus planning protocol: planning methods for creativity-centred design

Table 2.2 Resource accessibility and technological support in Ubiquitous Musical activities

N	Average age	Sex		Musical training	
10	35	Male	Female	Musicians	Nonmusicians
		3	7	5	5

systems may facilitate the access to informational resources through support mechanisms for social interaction. But in educational contexts, these same mechanisms may hinder creativity by limiting the type of relationships that agents establish among each other and by constraining the access to the available material resources. The purpose of our experimental studies is to identify the minimal requirements that would foster creativity, avoiding the introduction of unnecessary restrictions on the subjects' creative approaches (Fig. 2.2).

We conducted two workshops during July 2011. Teachers from a public high school in São José, SC, Brazil—encompassing different ages, areas of expertise and levels of training—participated in one of the workshops. Music teachers from the NGO Música e Cidadania (M&C), featuring both formally trained musicians (with undergraduate degrees in music teaching) and practising musicians without academic training, took part in the other workshop (Table 2.2). The workshops focused on aspects of sonic creation and compositional processes and the possibilities afforded by the use of everyday technologies for educational music making, emphasising the assessment of the complete creative experience as opposed to usage of isolated tools. In addition to enhancing the perception of our everyday listening environment, the workshops had two objectives: (1) to propose that teachers carry out their own interventions and creative experiences taking a standpoint of active engagement in authoring their own sound environments (*soundscape approach*) and (2) to encourage teachers to appropriate new technologies to explore the possibilities of sonic interventions, carrying out their own musical creations inspired by their daily activities (*eco-compositional approach*).

2.6.1 First Workshop: Procedures

The meetings were divided into two stages:

1. Low-tech (i.e. the use of technological resources was optional) featuring activities such as "ear cleaning" and "soundscape walks" [44, 47];
2. High-tech activities (with focus on the use of technological resources) based on eco-composition [21, 23] and Ubiquitous Music [25, 34].

During the workshops, these activities provided the context for a series of actions directed to:

(a) discuss concepts related to technology, media, sound, music and ubiquitous computing;
(b) handle and explore the possibilities of consumer media gadgets (portable computers, phones, mp3 players) and software tools (sound and musical data editors, sequencers and digital signal-processing tools);
(c) produce and create sounds and soundtracks through the combined use of technologies;
(d) replicate, evaluate and assess results through controlled experiments;
(e) share and discuss the creative experiences.

The first meeting (low tech) was performed with a group of mixed profiles, encompassing high school teachers with no musical training and music teachers from the M&C NGO who had formal training or previous ad hoc musical experience. On the first day, the participants were asked to perform ear cleaning activities, listing all the found sounds within the school environment. Both musicians and nonmusicians described sounds produced by human agents (screams, beatings, people talking and pushing chairs). During the activity, a section of the school was under construction, so there were sounds of machines, saws, drills and hammers mingling with the usual school sound environment. After providing descriptions of the sounds heard within the school premises, subjects were asked to take up a route from the school to another building. During this soundwalk, subjects wrote down what they heard and made recordings with their MP3 recorders and mobile telephones. Although some subjects were familiar with the use of portable audio devices, everyone—including the musicians—preferred to document their path by jotting down descriptions of the sounds on paper. A wide variety of sounds were listed. Some sounds were described as being surprising, while others were labelled as funny.

2.6.2 First Workshop: Results

During the low-tech session, there was a certain level of anxiety among the participants regarding what they were doing. Several subjects expressed that they had never done this type of activity before, and in several occasions, they interrupted their listening tasks with verbal observations and questions. After each listening session, the group discussed the experience. Overall, they considered the proposal to be fun. One of the younger teachers felt difficulty to focus on exploring the sonic aspects of the experience. This person did not enjoy doing the activity. Most of the M&C music teachers had previous exposure to the soundscape approach through readings during their music training. This group of subjects felt it was important to adopt open-ended activities and nontraditional methods for practical creative work with their students.

2.6.3 Second Workshop: Procedures

The second workshop (high tech) was carried out in the school's computer laboratory and outside the school premises. Before dealing with the tools, the participants discussed technical concepts related to musical data formats and open-source editors and sequencers. Recalling their soundscape field activities, teachers expressed their impressions on issues such as noise pollution and the environment. After this initial discussion, the coordinator (Maria Helena de Lima) exposed several ideas related to eco-composition and its relation to the creative application of technology. Participants brought up previous experiences with sound software and formulated questions regarding the tools they were going to use. At this point the sound editor Kristal was introduced, and sound samples were provided. Subjects were instructed to explore the sonic possibilities of the tool with the objective to obtain a musical product. Participants worked in groups of two and three members. The activities lasted 1 h. All groups made use of the sonic materials provided and managed to come up with a satisfactory musical product within the given timeframe.

2.6.4 Second Workshop: Results

Several subjects assessed the second experiment (high tech) as being difficult. Out of the seven participants, only three found it easy to deal with the tools provided. The exploration of materials and the creation of sonic products were described as fun experiences, and these participants enjoyed sharing their results. The younger teachers that had no formal music training frequently laughed at the results and used expressions like "that's cool" and "awesome!" while performing the tasks. One of the teachers (with extensive musical experience but no formal musical training) showed a conscious and concentrated effort to attain musically interesting products and evaluated the activity as being enjoyable. During the exploratory activity, the two teachers with formal music education—i.e., with previous experience in instrumental practice and expertise in traditional musical notation—had great difficulties in executing the task. These subjects were very concerned with instrumentally oriented musical parameters, such as pitch and rhythm, and were not able to explore other aspects of the sonic palette. Their focus on a specific type of result seemed to hinder their ability to test new sonic outcomes. As a whole, we could classify these results as a two-by-two matrix. Subjects with formal music training and subjects with unstructured musical experience had very different reactions to eco-compositional techniques. The first group did little exploration and had difficulties in realising creative activities. The second group described these activities as being easy, enjoyed the process and attained a high level of engagement. Naive subjects were less receptive than musicians to soundscape proposals. Despite the fact that the low-tech activities were not very demanding, they generated anxiety and estrangement among nonmusicians.

2.7 Phase 3: Prototyping

In order to test the applicability of the interaction metaphor *time tagging* in the context of a Ubiquitous Musical activity, we focused on one of the most basic uses of technology for sonic work: mixing. Mixing was chosen for three reasons. First, it encompasses a set of actions that can be isolated, quantified and analysed as a coherent whole without disrupting the broader context of music-making activities. Second, the actions executed in mixing can be supported by tools developed for consumer-level devices. Third, this activity affords a musically meaningful result: the sonic mix. By focusing on the temporal control of sounds, we can define mixing as a subclass of a general compositional problem related to the field of interaction. Because the methodological focus is aligned with the procedures adopted in sampling, this separation does not have a negative impact on the ecological validity of the process.

2.7.1 Definition of Mixing

Mixing constitutes an activity in which an agent or agents change the state of a set of sampled sounds from a random distribution to a temporally organised state. The original state can be described as a sample pool. The final state constitutes the mix.

2.7.2 Definition of Time Tagging

As an interaction metaphor, time tagging defines a process by which a set of unordered virtual elements or processes are layered onto an abstract one-dimensional structure—a tagged timeline (Fig. 2.3).

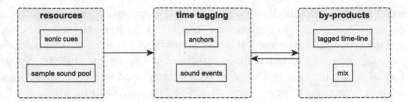

Fig. 2.3 Time-tagging interaction metaphor

2.7.3 MixDroid Prototype

The study involved the implementation of a prototype that supported a stripped-down definition of the activity at hand. By mapping a series of snapshots of a real-world sonic experience onto the constrained space of the mixing procedure, we obtained a metaphor that captures the minimum variables necessary for the study of mixing: time tagging [24]. The variables controlled in this interaction metaphor are time tags and sound samples. This simplified definition fulfilled two requirements. On one side, it was simple enough to provide the basis for an easily scalable interaction technique. On the other, it preserved the musical complexity of the process. MixDroid 1.0 is a testbed for the application of the time-tagging metaphor to mixing. In mixDroid, time tagging proceeds by synchronous sequential access of each of the sampled sounds stored in memory. Sound data is sent to a mixing buffer which constitutes the tagged timeline. While the user accesses the samples, the contents of the buffer are being played. This mechanism provides sonic cues that serve to shape the temporal anchoring process. The user listens to the sounds on the timeline, and the samples being played provide cues for the new sounds that will enter the mix. Finally, the new sound entering the timeline is tagged when the user presses a key. In time tagging, discrete physical elements of the interface are mapped onto the virtual elements or processes to be tagged. Our implementation features the mobile phone's touchscreen as our access infrastructure. A virtual keypad—comprised of nine soft buttons—is used to trigger the samples chosen by the user. Given that a numeric keypad—with up to 12 keys—is found on most mobile and fixed devices, this interaction metaphor can easily be translated to a multiplicity of low-cost gadgets.

2.7.4 Preliminary Experimental Results

Time-tagging evaluations yielded consistent high ratings for usability support both by musically experienced and untrained users [24, 41]. Keller and co-workers' [24] initial study showed that time tagging can be used as an interaction metaphor for mixing. At a minimum, time tagging reduces computational costs by eliminating the need for visually oriented support for mixing operations. This reduction is a key requirement for deployment on low-budget mobile platforms. Thus, time tagging stands better chances of success than graphics-based metaphors when it comes to supporting music making in the cloud.

2.8   Phase 4: Creativity Assessment of Time Tagging

After confirming the feasibility of the use of time tagging in ubiquitous mixing by an expert musician [24] and by novice users [41] within lab settings, we proceeded to a more detailed experimental analysis of its support in everyday contexts. To assess the impact of the location on pervasive musical activities, we designed an exploratory field study of the time-tagging metaphor encompassing four variables: musical training, activity type, location and type of sonic materials. This study provides an example of creativity assessment in everyday settings, complementing the results discussed in phase 2 (planning).

2.8.1   Procedures

Six subjects, musicians and nonmusicians, performed two types of creative musical activities using mixDroid: creation and imitation. The objective of the activity-labelled *creation* was to produce a sonic mix lasting approximately 30 s. *Imitation* engaged the users in listening and reproducing an already existing mix through a single trial. Two types of sound samples were employed: animal sounds (prominently frog sounds) and urban sounds (traffic sounds recorded at a busy road). A total of 47 iterations were done involving multiple locations: inside an isolated environment (studio) and at the sites where the sound samples had been gathered (street and pond).

Creativity ratings were obtained after each trial by applying the CSI-NAP protocol [27], an adapted version of the Creative Support Index proposed by Carroll et al. [6]. We translated the factors to simple direct questions in Portuguese. Qualitative descriptors were added to clarify the meanings of the evaluations. All questions were answered using a zero to ten likert-type scale. A field for comments was also made available.

2.8.2   Creativity Assessment Results

Overall results yielded high scores, particularly for the factors *enjoyment* and *collaboration*. No differences were found on the subjects' assessments of the activity type (*creation* vs. *imitation*). Both sound-sample classes yielded similar results. Activities carried out by musicians inside the studio got the lowest scores, but initially we could not determine whether this effect was correlated to the subject profile or to the location. A more refined analysis unveiled compound effects due to the type of sample and the location of the activity on the *explorability* factor (for animal sounds) and on *explorability, productivity* and *concentration* (for urban sounds).

The subjects' higher scores on the productivity factor for activities realised in external settings are baffling. From an eminent-creativity perspective, we would expect focused studio work to be more rewarding to musicians than other settings that are highly noise (street) or unfamiliar (pond). On a similar vein, novice participants should have yielded more consistent results when mixing inside the studio than at the outdoor environments. The tendency observed was the opposite. Ratings of the productivity factor were less consistent and lower for the studio condition.

2.8.3 Discussion of the Time-Tagging Study

The subjects' higher scores on the *productivity* factor for activities realised in external settings are baffling. From an eminent-creativity perspective, we would expect focused studio work to be more rewarding to musicians than other settings that are highly noise (street) or unfamiliar (pond). On a similar vein, novice participants should have yielded more consistent results when mixing inside the studio than at the outdoor environments. The tendency observed was the opposite. Ratings of the *productivity* factor were less consistent and lower for the studio condition.

Explorability was also consistently rated higher outside than inside the studio. This result supports the hypothesis that subjects are actually using environmental cues as proxies for the mixing activity. One factor that could undermine this conclusion would be high ratings for the *concentration* factor indicating that increased cognitive effort was necessary to execute the activity. This increase was apparent only for the street condition. But given the higher evaluation of *productivity*, more concentrated effort could also mean more *engagement* in the activity. Future experiments assessing engagement vs. attention level are necessary to settle this aspect of the analysis. In any case, the combination of higher *productivity* and *explorability* in outdoor settings points to a preliminary conclusion that time tagging provides support for enhanced creative on-site mixing experiences.

Conclusions

We described four approaches to creativity-centred design and provided examples of their application within the context of Ubiquitous Music practice. *Defining design strategies* involves methodological decisions that impact the experimental results and their interpretation. *Planning* involves experiments that provide initial data to determine basic requirements and unforeseen aspects of the creative experience. *Prototyping* entails materialisation of methodological proposals by technological means. Finally, creativity *assessment* techniques provide qualitative and quantitative results that can be used to re-evaluate the proposals put forth in the other stages.

Perspectives for Future Work

Through a discussion of a new definition of Ubiquitous Music, we unveiled two intersections with everyday creativity research: creative musical activities that do not demand expertise and activities that do not require specialised venues. Aspects of mobility and connectivity of ubiquitous systems were discussed in relation to human agents and material resources that afford musical activities through sustainable creativity support tools. We argued for investment of efforts on providing support for everyday creativity through creativity-centred design. Four methods were proposed: defining design strategies, planning, prototyping interaction and signal processing and creativity assessment. The UbiMus planning study yielded a set of social, material and procedural requirements. Based on these results, an interaction metaphor—time tagging—was applied to tackle audio mixing in ubiquitous contexts. The product was the prototype mixDroid. Six subjects participated in an exploratory field study using mixDroid, including two experimental conditions: inside the studio and in the locations where the sound samples were recorded. Subjects gave higher creativity scores to two factors—*explorability* and *productivity*—when activities were executed outside the studio. Future experiments targeting the influence of mobility on everyday creativity will need to unravel the relationship between cognitive effort and engagement during on-site creative activities. In any case, the results of the time-tagging study supported previous findings on the effectiveness of this metaphor for creative mixing. Together with the studies reported in [39] and [28], this is the first application of creativity-centred design within the realm of everyday creativity studies and Ubiquitous Music making.

References

1. Amabile, T.: Creativity in Context. Westview Press, Boulder (1996). http://books.google.com. br/books?id=hioVn_nl_OsC
2. Bennet, S.: The process of musical creation: interview with eight composers. J. Res. Music Educ. **24**, 3–13 (1976)
3. Bevan, N.: Measuring usability as quality of uses. Software Qual. J. **4**, 115–150 (1995)
4. Brinkmann, P.: Making Musical Apps: Using the Libpd Sound Engine. O'Reilly & Associates Incorporated, Sebastopol (2012). http://books.google.com.br/books?id=6uUNZ9yH7wsC
5. Burnard, P., Younker, B.A.: Problem-solving and creativity: insights from students? Individual composing pathways. Int. J. Music Educ. **22**, 59–76 (2004)
6. Carroll, E.A., Latulipe, C., Fung, R., Terry, M.: Creativity factor evaluation: towards a standardized survey metric for creativity support. In: Proceedings of ACM Creativity & Cognition, Berkeley (2009)
7. Chen, C.W.: A synthesis process model of creative thinking in music composition: visual images and music. Ph.D. thesis, Royal Melbourne Institute of Technology (2006)
8. Collins, D.: A synthesis process model of creative thinking in music composition. Psychol. Music **33**(2), 193–216 (2005)
9. Costa, C.A., Yamin, A.C., Geyer, C.F.R.: Toward a general software infrastructure for ubiquitous computing. IEEE Pervasive Comput. **7**(1), 64–73 (2008)
10. Dingwall, C.: Rational and intuitive approaches to music composition: the impact of individual differences in thinking/learning styles on compositional processes. Ph.D. thesis, University of Sydney (2008)
11. Essl, G., Rohs, M.: Interactivity for mobile music-making. Organised Sound **14**(2), 197–207 (2009). doi:10.1017/S1355771809000302. http://dx.doi.org/10.1017/S1355771809000302
12. Ferraz, S., Keller, D.: Mdf: Proposta preliminar do modelo dentro-fora de criação coletiva. Cadernos de Informática **8**(2), 57–67 (2012)
13. Fitzmaurice, G.W., Ishii, H., Buxton, W.: Bricks: laying the foundations for graspable user interfaces. In: Proceedings of the ACM SIGCHI Conference on Human Factors in Computing Systems, pp. 442–449 (1995)
14. Flores, L.V., Pimenta, M.S., Keller, D.: Patterns for the design of musical interaction with everyday mobile devices. In: Proceedings of the 9th Brazilian Symposium on Human Factors in Computing Systems. Belo Horizonte, Brazil (2010)
15. Flores, L.V., Pimenta, M.S., Keller, D.: Patterns of musical interaction with computing devices. In: Proceedings of the III Ubiquitous Music Workshop (III UbiMus). Ubiquitous Music Group (g-ubimus), São Paulo, Brazil (2012). http://compmus.ime.usp.br/ubimus/pt-br/node/33
16. Gould, J., Conti, J., Hovanvecz, T.: Composing letters with a simulated listening typewriter. Commun. ACM **26**(4), 295–308 (1983)
17. Hickey, M.: Creative thinking in the context of music composition. In: Hickey, M. (ed.) Why and How to Teach Music Composition: A New Horizon for Music Education, pp. 31–53. The National Association for Music Education, Reston (2003)
18. Hornbaek, K.: Current practice in measuring usability: challenges to usability studies and research. Int. J. Hum. Comput. Stud. **64**(2), 79–102 (2006). doi:10.1016/j.ijhcs.2005.06. 002. http://www.cse.chalmers.se/research/group/idc/ituniv/kurser/09/hcd/literatures/Hornbaek %202006%20usability%20measurement%20methods.pdf
19. Huang, E.M., Truong, K.N.: Breaking the disposable technology paradigm: opportunities for sustainable interaction design for mobile phones. In: Proceedings of the SIGCHI Conference on Human Factors in Computing Systems, pp. 323–332 (2008)
20. Hutchins, E.: Cognition in the Wild. MIT Press, Cambridge (1995). http://www.amazon.com/ exec/obidos/redirect?tag=citeulike07-20&path=ASIN/0262082314
21. Keller, D.: Compositional processes from an ecological perspective. Leonardo Music J. **10**, 55–60 (2000)
22. Keller, D.: Paititi: a multimodal journey to el dorado. Ph.D. thesis, Stanford University (2004)

23. Keller, D., Capasso, A.: New concepts and techniques in eco-composition. Organised Sound **11**(1), 55–62 (2006)
24. Keller, D., Barreiro, D.L., Queiroz, M., Pimenta, M.S.: Anchoring in ubiquitous musical activities. In: Proceedings of the International Computer Music Conference, pp. 319–326. MPublishing, University of Michigan Library, Ann Arbor (2010). http://hdl.handle.net/2027/spo.bbp2372.2010.064
25. Keller, D., Flores, L.V., Pimenta, M.S., Capasso, A., Tinajero, P.: Convergent trends toward ubiquitous music. J. New Music Res. **40**(3), 265–276 (2011). doi:10.1080/09298215.2011.594514. http://www.tandfonline.com/doi/abs/10.1080/09298215.2011.594514
26. Keller, D., Lima, M.H., Pimenta, M.S., Queiroz, M.: Assessing musical creativity: material, procedural and contextual dimensions. In: Proceedings of the National Association of Music Research and Post-Graduation Congress—ANPPOM (Anais do Congresso da Associação Nacional de Pesquisa e Pós-Graduação em Música—ANPPOM), pp. 708–714. National Association of Music Research and Post-Graduation (ANPPOM), ANPPOM, Uberlândia (2011). http://www.anppom.com.br/anais.php
27. Keller, D., Pinheiro da Silva, F., Giorni, B., Pimenta, M.S., Queiroz, M.: Spatial tagging: an exploratory study (marcação espacial: estudo exploratório). In: Proceedings of the 13th Brazilian Symposium on Computer Music. SBC, Vitória (2011). http://compmus.ime.usp.br/sbcm/2011/
28. Keller, D., Ferreira da Silva, E., Pinheiro da Silva, F., Lima, M.H., Pimenta, M.S., Lazzarini, V.: Everyday musical creativity: an exploratory study with vocal percussion (criatividade musical cotidiana: Um estudo exploratório com sons vocais percussivos). In: Proceedings of the National Association of Music Research and Post-Graduation Congress—ANPPOM. ANPPOM, Natal (2013)
29. Lave, J., Wenger, E.: Situated Learning: Legitimate Peripheral Participation. Cambridge University Press, Cambridge (1991)
30. Lazzarini, V.: A toolkit for audio and music applications in the xo computer. In: Proceedings of ICMC 2008, pp. 62–65. Belfast, Northern Ireland (2012)
31. Lazzarini, V., Yi, S., Timoney, J., Keller, D., Pimenta, M.: The mobile Csound platform. In: Proceedings of ICMC 2012 (2012)
32. Lima, M.H., Keller, D., Pimenta, M.S., Lazzarini, V., Miletto, E.M.: Creativity-centred design for ubiquitous musical activities: two case studies. J. Music Technol. Educ. **5**(2), 195–222 (2012). doi:10.1386/jmte.5.2.195_1. http://www.ingentaconnect.com/content/intellect/jmte/2012/00000005/00000002/art00008
33. Miletto, E.M., Flores, L.V., Pimenta, M.S., J.R., Santagada, L.: Interfaces for musical activities and interfaces for musicians are not the same: the case for codes, a web-based environment for cooperative music prototyping. In: 9th international conference on multimodal interfaces, pp. 201–207. ACM, Nagoya (2007)
34. Miletto, E.M., Pimenta, M.S., Bouchet, F., Sansonnet, J.P., Keller, D.: Principles for music creation by novices in networked music environments. J. New Music Res. **40**(3), 205–216 (2011). doi:10.1080/09298215.2011.603832. http://www.tandfonline.com/doi/abs/10.1080/09298215.2011.603832
35. Mitchell, W.J., Inouye, A.S., Blumenthal, M.S.: Beyond Productivity: Information, Technology, Innovation, and Creativity. The National Academies Press, Washington (2003). http://www.nap.edu/openbook.php?record_id=10671
36. Mumford, M.D., Hester, K., Robledo, I.: Methods in creativity research: Multiple approaches, multiple methods. In: Mumford, M.D. (ed.) Handbook of Organizational Creativity, pp. 39–64. Elsevier Science, Waltham (2011)
37. Pimenta, M.S., Flores, L.V., Capasso, A., Tinajero, P., Keller, D.: Ubiquitous music: concepts and metaphors. In: Proceedings of the XII Brazilian Symposium on Computer Music, pp. 139–150. Recife, Brazil (2009)
38. Pimenta, M.S., Miletto, E.M., Keller, D., Flores, L.V.: Technological support for online communities focusing on music creation: adopting collaboration, flexibility and multiculturality from brazilian creativity styles. In: Azab, N.A. (ed.) Cases on Web 2.0 in Devel-

oping Countries: Studies on Implementation, Application and Use, Chap. 11. IGI Global Press, Vancouver (2012). doi:10.4018/978-1-4666-2515-0.ch011. http://www.igi-global.com/chapter/technological-support-online-communities-focusing/73062

39. Pinheiro da Silva, F., Keller, D., Ferreira da Silva, E., Pimenta, M.S., Lazzarini, V.: Everyday musical creativity: exploratory study of ubiquitous musical activities (criatividade musical cotidiana: estudo exploratório de atividades musicais ubíquas). Música Hodie **13**, 64–79 (2013). http://www.musicahodie.mus.br/13.1/Artigo_Cientifico_05.pdf

40. Plucker, J.A., Makel, M.C.: Assessment of creativity. In: Kaufman, J.C., Sternberg, R.J. (eds.) The Cambridge Handbook of Creativity, pp. 20–47. Cambridge University Press, Cambridge (2010)

41. Radanovitsck, E.A.A., Keller, D., Flores, L.V., Pimenta, M.S., Queiroz, M.: mixdroid: Time tagging for creative activities (mixdroid: Marcação temporal para atividades criativas). In: Proceedings of the XIII Brazilian Symposium on Computer Music (SBCM). SBC, Vitória (2011). http://compmus.ime.usp.br/sbcm/2011

42. Richards, R.: Everyday creativity and new views of human nature: psychological, social, and spiritual perspectives. Am. Psychol. Assoc. (2007). http://books.google.ie/books?id=gBHcngEACAAJ

43. Richards, R., Kinney, D.K., Benet, M., Merzel, A.P.: Assessing everyday creativity: characteristics of the lifetime creativity scales and validation with three large samples. J. Pers. Soc. Psychol. **54**, 476–485 (1988). doi:10.1037//0022-3514.54.3.476

44. Schafer, R.M.: The Tuning of the World. Knopf, New York (1977)

45. Shneiderman, B., Fischer, G., Czerwinski, M., Resnick, M., Myers, B.A., Candy, L., Edmonds, E.A., Eisenberg, M., Giaccardi, E., Hewett, T.T., Jennings, P., Kules, B., Nakakoji, K., Nunamaker, J.F., Pausch, R.F., Selker, T., Sylvan, E., Terry, M.A.: Creativity support tools: report from a U.S. national science foundation sponsored workshop. Int. J. Hum. Comput. Interact. **20**(2), 61–77 (2006). http://dblp.uni-trier.de/db/journals/ijhci/ijhci20.html#ShneidermanFCRMCEEGHJKNNPSST06

46. Tanaka, A.: Mobile music making. In: Proceedings of the International Conference on New Interfaces for Musical Expression, pp. 154–156 (2004). http://www.nime.org/proceedings/2004/nime2004_154.pdf

47. Truax, B.: Genres and techniques of soundscape composition as developed at simon fraser university. Organised Sound **7**(1), 5–14 (2002). doi:10.1017/S1355771802001024. http://dx.doi.org/10.1017/S1355771802001024

48. Weinberg, G.: Interconnected musical networks: toward a theoretical framework. Comput. Music J. **29**(2), 23–39 (2005)

49. Weiser, M.: The computer for the 21st century. In: Baecker, R.M., Grudin, J., Buxton, W.A.S., Greenberg, S. (eds.) Human-computer Interaction, pp. 933–940. Morgan Kaufmann Publishers, San Francisco (1995). http://dl.acm.org/citation.cfm?id=212925.213017

50. Wenger, E.: Communities of practice and social learning systems: the career of a concept. In: Blackmore, C. (ed.) Social Learning Systems and Communities of Practice. Springer and The Open University, London (2010)

Part II
Applications

Chapter 3
Repertoire Remix in the Context of *Festival City*

Akito van Troyer

Abstract Repertoire Remix enables remote audience members to dynamically suggest their musical preferences for live web-streaming musical improvisation sessions. The semantic web interface encourages remote participants to collaboratively use "stirring" mouse gestures to influence the size of graphical bubbles that contain composers' names. The accumulated weight is then interpreted by musicians to improvise. This chapter documents the first pilot run of the Repertoire Remix system, explores challenges in designing a real-time shared music style arranging system for networked live improvisation and interprets the resulting performance by assessing the participants' mouse gestures collected during the pilot run.

3.1 Repertoire Remix in the Context of *Festival City*

Festival City is a symphonic work commissioned by the Edinburgh International Festival (EIF) that premiered in August 2013 [7]. The main objective of this work was to have a composer, Tod Machover, and people in the Scottish capital, Edinburgh, to collaborate in composing a final piece of music in one of the concert series taking place in EIF. In realising this objective, our first step was to develop a number of web-based music composition applications that enabled people from Edinburgh to remotely and asynchronously collaborate in music making on the web [22]. These composition applications brought together an unprecedented number of people from diverse backgrounds to contribute in pursuing our main objective and also established a unique model for creating intricate collaborations between experts and nonexperts in composing music.

While these web-based composition environments achieved our main objective, we wanted to further experiment with the *Festival City* project to engage our audience more intimately by providing them opportunities to remotely interact with the composer and musicians in a live context. To accomplish this goal, we created Repertoire Remix, a web browser-based application that can creatively engage remote audience members to become part of the real-time improvisational

A. van Troyer (✉)
MIT Media Lab, Cambridge, MA, USA
e-mail: akito@media.mit.edu

© Springer International Publishing Switzerland 2014   51
D. Keller et al. (eds.), *Ubiquitous Music*, Computational Music Science,
DOI 10.1007/978-3-319-11152-0_3

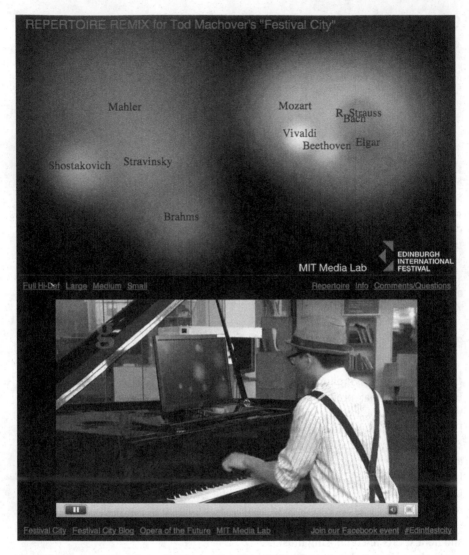

Fig. 3.1 A screenshot of the Repertoire Remix web interface during the pilot run of the improvisation session

performance process [12]. Repertoire Remix does not require website visitors to have specialised musical training. It uses a visual interface consisting of ten morphing graphical bubbles that represent famous composers and their music styles (see Fig. 3.1), and website visitors simply "stir" their mouse in the interface to suggest their musical preference to performers. The composer bubbles then change size according to the mouse gestures, which then further influence how the performers improvise. While remote participants influence the composer bubbles, the director, a person who oversees and moderates the entire performance session, controls a second interface with sliders and buttons that determines how the composer bubbles overlap, bounce against each other or fuse together. The aim of having a director is to give the performers more ideas about how to combine and transform their improvisation.

Repertoire Remix is also a system that collects various real-time data including MIDI data events, video, audio and remote participants' mouse gestures. We collected these data to facilitate our original objective to utilise such data in the final piece of music that can be performed at EIF. For our first pilot run, the solo piano improvisation session, performed by Tae Kim and directed by Tod Machover, was recorded and later incorporated into the final piece of music that then got performed at EIF [17].

3.2 Audience Integration in Network Music Environments

Many network music projects incorporate audience members as integral parts of their performances. Such performances typically encourage them to be expressive and collaboratively influence the performance in real time [16]. The earliest examples of such performances are *Brain Opera* and *Cathedral*, where online collaborative interactions remotely influenced live concert performances [6, 15]. These projects provided opportunities to rethink the nature of collective interaction in public spaces when remote participants were also involved in this process. Similarly, more recent projects such as Quintet.net [13] and Graph Theory [8] provide feedback and music-making systems for remote participants to contribute to performances in real time. These projects demonstrate the capability of network music performance systems to provide unique and enriching interactive musical experiences to remote audience members.

Recent developments in computer network systems have led to new approaches for composition and improvisation for musical novices. These network music systems, termed "shared sonic environments" [2], enable remote participants to collaboratively create shared soundscapes. Such environments can often blur the boundaries between audience and performers when, for example, remote participants become the central agents in producing sounds that are then mixed and processed in real time over a broadcast/network system to create a musical performance [11]. In other cases, remote participants' collaborative activities on a simple visual-oriented web interface result in creating unique music improvisations and compositions [4, 5]. These projects partly rely on visual information to show other participants' activities in the shared environment. In addition, this visual information often works as a feedback system and helps participants by suggesting other people's presence in the environment. Such use of visual information also facilitates competition and collaboration among the participants, making musical creation fun and social.

In designing Repertoire Remix, we also explored how existing real-time music notation and audience participation-based performance systems facilitate interaction between audience and performers to the point that they can both contribute to creating an improvised performance. For this interaction to happen, we imposed a condition that such performance systems do not require participants to learn about or be skilled at operating the interface, as most of them are likely to be first-time users and have no time for a learning curve [3]. Similar approaches can be seen in many of the prior real-time music notation systems. Two examples are real-time music notation systems in which the audience was able to shape the ongoing music at the performance site [1, 9] and online music composition systems that enable website visitors to compose music through an open-form music scoring method that was then used for the actual performance [10]. The challenge of these projects was to enhance the participatory experience of musical novices through simple music composition interfaces. Furthermore, these projects gave new perspectives to the prospect of collaboration among amateur and professional musicians.

3.3 Design and Implementation

The core interaction flow of Repertoire Remix is illustrated in Fig. 3.2. The figure can be divided into three parts: remote locations, servers and headquarters. Remote locations represent the website visitors. They use a web browser-based interface that is designed to encourage them to stir their mouse to influence the bubbles within the shared music style arranging environment (SMSAE) (see the top portion of Fig. 3.1). At the same time, they also see and hear the musicians' performance and director's musical instructions through web-streaming video at the bottom part of the interface (see the bottom portion of Fig. 3.1). These activities for website visitors are made possible by the servers that relay interaction data and web-streaming video to the remote client. Servers also make the remote collaboration happen between

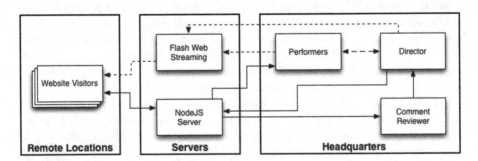

Fig. 3.2 The core interaction flow in Repertoire Remix. *Bold lines* represent interaction data flow, while *dotted lines* show the performance audio/video capturing and streaming. The *dashed line* represent a verbal dialog that can take place between director and performers during the session

remote locations and headquarters. Headquarters is where the actual performance is taking place, and the Repertoire Remix system assumes that performers, director and comment reviewers exist in the same physical space. Performers improvise music during the session using a simplified version of the SMSAE interface while a director, who has his/her own unique web interface, oversees and moderates the interaction between performer and remote audience. Although optional, comment reviewers exist to filter out unwanted comments from remote audience and deliver appropriate comments to the performers and the director for live responses via the web stream. Comment reviewers also have their own unique interface to regulate comments. Repertoire Remix consists of four different interfaces, but the system is able to realise bidirectional communication among them through the SMSAE, video streaming and web comments.

3.3.1   Shared Music Style Arranging Environment

One of our goals in designing the Repertoire Remix system was to provide a shared visual environment between participants, who may have little or no music training, and professional musicians, so that they can communicate their musical ideas and intentions through the same abstract visual representations. We chose to illustrate such abstract visual representation using graphical bubbles that contain composer names. These bubbles can morph their texture, size and positions during the course of the improvisation session. We believe that this approach can empower novices to easily express their musical ideas only using their mouse gestures. The system works so that whenever the participants stir their mouse, the closest composer bubbles grow in size. This approach of using abstract visual representation with simple mouse gestures also helped promote our ideas to keep the musicians' interpretation of music style suggestions open. They do so by viewing a simplified noninteractive version of the SMSAE interface responding to aggregated user data shown in Fig. 3.1 with a clock to keep track of time.

To technically realise SMSAE, we used NodeJS as the underlining server architecture. NodeJS is a server-side platform for building fast and scalable web applications [20]. It is also simple and very well suited for creating servers that can hold shared data across all clients. The type of data shared across all browsers were related to the properties of the composer bubbles and the logical propositions of who influenced which of those bubbles. The shared data are then communicated among the clients using NowJS which makes building real-time web applications using WebSockets effortless within the NodeJS environment [21, 23]. The data can also be stored on the server using Mongoose, a NodeJS module that interfaces with MongoDB [18, 19], for an optional post analysis of a performance. Client browsers that support HTML 5 standards are then used to render the graphical bubbles using the canvas element [14].

Fig. 3.3 A slider and button interface for a director

3.3.2 Slider and Button Interface for Director

In designing the Repertoire Remix performance system, we felt the need for a person who can moderate and run the improvisation session since our first pilot run was not only going to involve several tryout performances within an hour but also a discussion between remote participants and the performers. For this reason, we created a role of director in the system who could facilitate the dialog and also oversee the performance sessions. The director has his/her own unique interface with sliders and buttons for influencing some parameters of the SMSAE (see Fig. 3.3). These parameters are related to the properties of the composer bubbles such as their position, blurring effect, size growth rate and speed. The sliders for growth rate and the buttons at the bottom of the interface are ways for the director to take control over composer bubbles that are getting "out of control" so that the improvisation session can proceed with appropriate variations and some cohesiveness. Other sliders, such as bubble position and blurring effect, are used to give performers variations in how to combine different styles of music. As the nature of such an abstract/nontraditional interface can be confusing for the remote participants in understanding the interaction, for our first pilot run, we explained the interface functionality to remote participants prior to and at the beginning of the session so that they will not be confused by the unexpected behaviour of the composer bubbles during the performance.

3.3.3 Web Streaming

The setup for the Adobe Flash-based web-streaming system is entirely independent from the rest of the interactive Repertoire Remix system. This enables anyone using the Repertoire Remix system to flexibly arrange how to capture the audio/visual contents and how to send the streaming view to the remote participants. In our first pilot run, for instance, we operated two cameras to capture the improvisation session from different view angles, which then were mixed live using a video mixer. In addition, we also used an audio mixer to capture several different sources of sounds from microphones (mostly for conversation) and MIDI piano sounds captured from Yamaha AvantGrand N3 piano. The MIDI events from the piano were ideal in our situation to synthesise electronic piano sounds and keep the quality of the audio at the maximal level for the web-streaming. The web streaming display at the bottom of the main interface served as the primary way for the remote participants to be able to view the performance in real time and see the effects of their participation in the performance.

3.3.4 Comment/Feedback System

Embedded in the main interface, the Repertoire Remix system also features a way for the remote participants to send live comments. When a participant sends a comment, the message is displayed in the interface illustrated in Fig. 3.4. Although this interface is not necessary for all performance scenarios, we intend this live commenting system to be interpreted by someone other than the director or the performers for filtering unwanted comments and delivering the right ones that promote conversation about the performance and how to improve it. The reason why

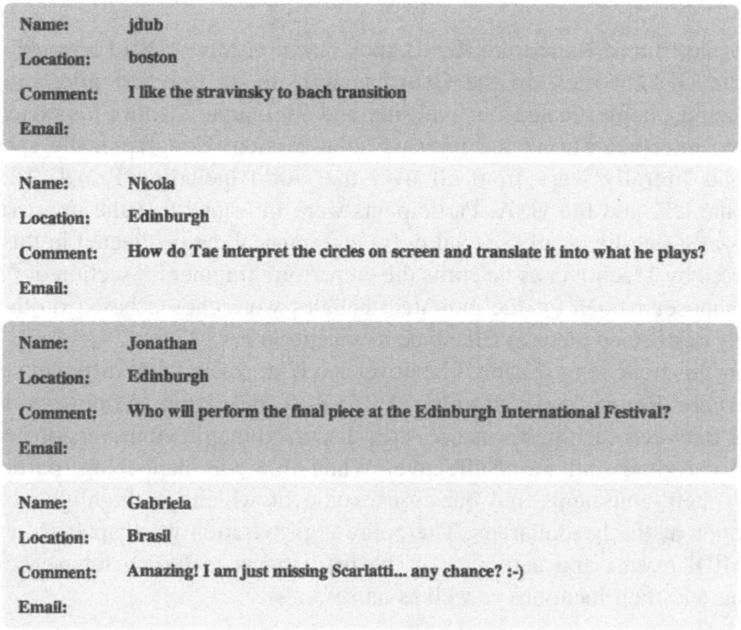

Name:	jdub
Location:	boston
Comment:	I like the stravinsky to bach transition
Email:	

Name:	Nicola
Location:	Edinburgh
Comment:	How do Tae interpret the circles on screen and translate it into what he plays?
Email:	

Name:	Jonathan
Location:	Edinburgh
Comment:	Who will perform the final piece at the Edinburgh International Festival?
Email:	

Name:	Gabriela
Location:	Brasil
Comment:	Amazing! I am just missing Scarlatti... any chance? :-)
Email:	

Fig. 3.4 An example interface for comment interpreters

we implemented such system was to enable remote participants to exchange their musical ideas that would otherwise be hard to communicate with the director and the performers through visual information alone. The comment system can enable high-level and sophisticated ways of sharing musical idea across remote locations.

3.4 Interpretation of Shared Visual Score by Performers

Our basic protocol for how the performers interpret the shared visual representation is based on the accumulated weight of the bubble sizes; the performers establish their improvisation style in favour of the bubbles that are bigger in size than others and build up the aggregated musical styles all together. In our first pilot run, for example, the pianist often took three to four bubbles that are larger than the rest and mixed the style sequentially in improvisational fashion. We also imposed that interpreting the position and speed of the composer bubbles will affect how chaotic or random the mix of styles becomes, while the blurring effect of each bubble will be used to influence how performers blend different music styles into one coherent performance style.

3.5 Pilot Run

The first pilot run of Repertoire Remix took place on July 9, 2013 between 7 p.m. and 8 p.m. GMT hosted on the Guardian website. In our first pilot run, over one thousand people visited our website, and 71 unique visitors interacted with the shared interface during the improvisation session. Geographically speaking, participants literally were from all over the world including Brazil, Thailand, Ireland, the UK and the USA. Participants were informed that the improvisation session was a one-time-only special event and musical data collected in this event will be used by Machover as he crafts the "repertoire fragments" section of *Festival City*. Composer names for the graphical bubbles were chosen based on the most frequently performed piece at EIF since its inception in 1947.

During this hour-long session, Machover and Kim created four different versions of Repertoire Remix improvisation, of 15, 8, 5 and 3 min, refining with each iteration. Between each performance, they discussed the previous version and how to make it tighter and more effective. While this was happening, participants also sent their comments and questions, some of which got highlighted in the conversation at the headquarters. The entire improvisation was captured in video, audio, MIDI events and activities of the participants including mouse gestures, comments and their locations as well as names.

3.6 Activity Assessment of the Pilot Run

This section explores the visualisation and analysis of the "mouse gesture" data logged by the system during the improvisation session. The data being analysed in this section were collected every time remote participants influenced one of the composer bubbles in the shared visual environment. Through the collected data, we looked for any identifiable and meaningful patterns that could provide quantitative information about the improvisation session and the behaviours of remote participants.

Figure 3.5 shows the aggregated mouse gestural influences of all remote participants on the graphical bubbles throughout the entire improvisation session. We counted all participants' mouse gestural influence on the bubbles every 12 s and visualised them to see when participants actually interacted with the bubble interface. The graph also indicates with grey regions when the test session and each improvisation session happened. The trend in the graph tells us that the increase in mouse activities happens consistently with times each improvisation sessions was taking place. In contrast to our worries about network delay and providing web-streaming content on time, the graph indicates that remote participants were able

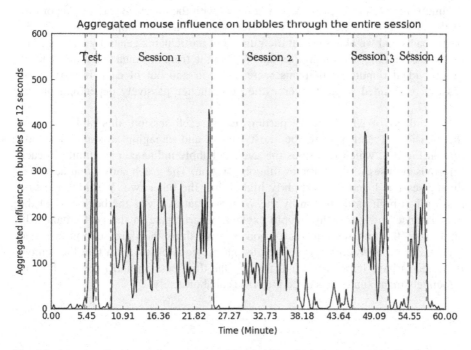

Fig. 3.5 Aggregated mouse influence on graphical bubbles throughout the entire improvisation session

Table 3.1 Summary and comparison of each improvisation session

	Session 1	Session 2	Session 3	Session 4
Performance time (s)	928	485	322	178
Number of participants	29	26	26	19
Number of influence	11,063	4,420	4,316	2,301
Average number of influence	381.48	170	166	121.1
Average number of influence per minute	20.04	15.53	19.71	9.14

to participate in influencing each improvisation session in real time with little lag between video and interaction. On the other hand, we also observe relatively intense activity in the intermissions between sessions 2 and 3 as well as sessions 3 and 4. This trend may suggest that the participants were becoming comfortable using the shared interface and were gaining intuition on the effect of the system on the performance.

Table 3.1 summarises the activity of the remote participants with respect to the duration, number of participants, number of influences on bubbles, average number of influences on bubbles and number of influences on bubbles per minute in each improvisation session. We mentioned in the previous section that the total number of unique remote participants, who interacted with the interface was 71. According to the table, however, each improvisation session had only approximately 20–30 participants and we also see that the number of participants generally decreased as the web-streaming session progressed. We assume from this data that a fair amount of individual remote participants were either in and out of each improvisation session or decided to just observe the performance passively depending on the session.

Regardless of who actually participated in each session, this table suggests that sessions 1 and 3 were the most fruitful and engaging sessions. The box plot in Fig. 3.6, which describes the average bubble influence per minute in each improvisation session, reinforces this observation. The graph shows that activities during session 1 and 3 are slightly higher than the other two. We think the first session had high participation by the visitors because it was the beginning of the performance and got many people excited to participate in the performance. In contrast to the first session, the reason why we think the third session also had high levels of interactivity is related to our previous observation that the remote participants have learned the interface by the third session and knowing how to effectively manipulate the shared interface collaboratively.

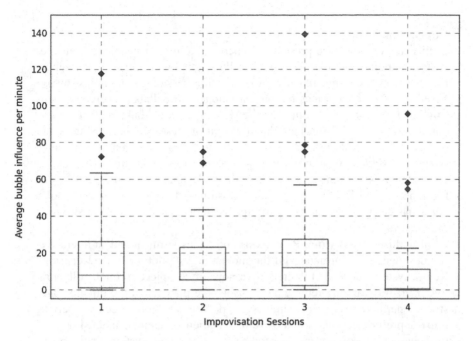

Fig. 3.6 A box plot of average bubble influence per minute in each improvisation session

3.7   Comments from Participants During the Pilot Run

We collected approximately 30 comments during the live improvisation session. The collected comments ranged from appraisal of the performance, questions on technical implementation of the system and also how they like to shape the music for the pianist. One of most common comments collected from participants was regarding the interpretation of the visual representation by the pianist. One participant asked "How does Tae interpret the circles on screen and translate it into what he plays?" We had explained how the improvisation session works through the live video stream and also in the instructions on the webpage that the pianist reads the accumulated weight of the composer bubble size to guide his improvisation, but the amount of comments we received shows that our effort in having a mutual understanding in the collaborative music creation between the pianist and remote participants did not always perfectly work out. We also received fair amount of comments that we sought from remote participants including comments about how to improve the next improvisation session. These comments include: "I like the slower start. It's a nice way to ease into it", "I like the Stravinsky to Bach transition", "More Stravinsky than just the Rite chord!" and "What sort of juxtapositions in particular really evoke Edinburgh for you?" These comments suggest to us that some remote participants were truly interacting with the interface to influence the course of improvisation with the pianist and the director.

Conclusion

In this chapter, we have presented Repertoire Remix, a system that enables remote audience members to dynamically suggest their musical preferences for live web-streaming musical improvisation sessions. The web browser-based interface encourages remote participants to collaboratively use "stirring" mouse gestures to influence the size of graphical bubbles that contain composers' names. The accumulated weight of these bubble sizes and other properties of the bubbles, such as position and texture, are then interpreted by talented musicians to improvise in a certain way in real time. The chapter also addresses a number of challenges in designing a real-time SMSAE for a networked live improvisational session and interprets the result of a performance by assessing the participants' mouse gestures collected during the pilot run.

In summary, Repertoire Remix was both an experiment and an interactive remote musical performance. Furthermore, the first pilot run of the Repertoire Remix system succeeded in creating a novel music piece through collaborative interaction among remote participants, the pianist and the director. This study generated new insights into Network Music Performance by receiving musical preferences suggested by remote audience members and using those to create an improvisational session responding to the system status in real time.

Acknowledgements The author would like to thank Muriel R. Cooper Professor Tod Machover, Simone Ovsey and Ben Bloomberg for their help in the creation of the Repertoire Remix web application. The author would also like to thank Tae Kim for his collaboration in making the pilot improvisation session a successful one. Thanks to the members of the Opera of the Future group at MIT Media Lab for your moral support.

References

1. Baird, K.C.: Real-time generation of music notation via audience interaction using python and gnu lilypond. In: Proceedings of the 2005 Conference on New Interfaces for Musical Expression, pp. 240–241, National University of Singapore, Singapore (2005)
2. Barbosa, Á.: Displaced soundscapes: a survey of network systems for music and sonic art creation. Leonardo Music J. **13**, 53–59 (2003)
3. Blaine, T., Fels, S.: Contexts of collaborative musical experiences. In: Proceedings of the 2003 Conference on New Interfaces for Musical Expression, NIME '03, pp. 129–134, National University of Singapore, Singapore (2003). http://dl.acm.org/citation.cfm?id=1085714.1085745
4. Bryan-Kinns, N., Healey, P.G., Leach, J.: Exploring mutual engagement in creative collaborations. In: Proceedings of the 6th ACM SIGCHI Conference on Creativity & Cognition, pp. 223–232. ACM, New York (2007)
5. Burk, P.: Jammin'on the web–a new client/server architecture for multi-user musical performance. In: ICMC 2000, Citeseer (2000)

6. Duckworth, W.: Making music on the web. Leonardo Music J. **9**, 13–17 (1999)
7. Edinburgh International Festival: *Festival City*. http://www.eif.co.uk/festival-city (2013)
8. Freeman, J.: Graph theory: interfacing audiences into the compositional process. In: Proceedings of the 7th International Conference on New Interfaces for Musical Expression, pp. 260–263. ACM, New York (2007)
9. Freeman, J.: Extreme sight-reading, mediated expression, and audience participation: real-time music notation in live performance. Comput. Music J. **32**(3), 25–41 (2008). doi:10.1162/comj.2008.32.3.25. http://dx.doi.org/10.1162/comj.2008.32.3.25
10. Freeman, J.: Web-based collaboration, live musical performance and open-form scores. Int. J. Perform. Arts Digit. Media **6**(2), 149–170 (2010). doi:doi:10.1386/padm.6.2.149_1. http://www.ingentaconnect.com/content/intellect/padm/2010/00000006/00000002/art00002
11. Freeman, J., Varnik, K., Ramakrishnan, C., Neuhaus, M., Burk, P., Birchfield, D.: Auracle: a voice-controlled, networked sound instrument. Organised Sound **10**(3), 221 (2005)
12. Guardian: Repertoire Remix. http://www.guardian.co.uk/music/interactive/2013/jul/05/tod-machover-festival-city-interactive-live-stream?INTCMP=SRCH (2013)
13. Hajdu, G.: Quintet. net–a quintet on the internet. In: Proceedings of the International Computer Music Conference, vol. 29, pp. 315–318 (2003)
14. HTML5 Canvas: http://www.w3.org/TR/2009/WD-html5-20090825/the-canvas-element.html (2009)
15. Machover, T.: Brain Opera. http://brainop.media.mit.edu (1996)
16. Miletto, E.M., Pimenta, M.S., Bouchet, F., Sansonnet, J.P., Keller, D.: Principles for music creation by novices in networked music environments. J. New Music Res. **40**(3), 205–216 (2011)
17. MIT Media Lab: Repertoire Remix: The Sounds of Edinburgh. http://blog.media.mit.edu/2013/08/repertoire-remix-sounds-of-edinburgh.html (2013)
18. MongoDB: http://www.mongodb.org/ (2011)
19. Mongoose: http://mongoosejs.com/ (2011)
20. NodeJS: http://nodejs.org/ (2011)
21. NowJS: https://github.com/Flotype/now (2011)
22. van Troyer, A.: *Festival City* Online Score Making Environments. http://edinburgh.media.mit.edu/scores/ (2013)
23. WebSocket: https://tools.ietf.org/html/rfc6455 (2011)

Chapter 4
Making Meaningful Musical Experiences Accessible Using the iPad

Andrew R. Brown, Donald Stewart, Amber Hansen, and Alanna Stewart

Abstract In this chapter we report on our experiences using ubiquitous computing devices to introduce music-based creative activities into an Australian school. The use of music applications on mobile tablet computers (iPads) made these activities accessible to students with a limited musical background and practicable in a general-purpose classroom setting. The activities were designed to be meaningful and contribute toward personal resilience in the students. We describe the approach to meeting these objectives and discuss results of the project. The chapter includes an overview of the ongoing project including its aims, objectives and utilisation of mobile technologies and software with generative and networkable capabilities. Two theoretical frameworks informed the research design: the meaningful engagement matrix and personal resilience. We describe these frameworks and how they inform the activity planning. We report on the activities undertaken to date and share results from questionnaires, interviews, musical outcomes and observation.

4.1 Introduction

This project builds on the authors' previous work with network music jamming systems [3] and youth resilience[13]. These research threads have come together in this project. Taking advantage of the ubiquitous nature of mobile computing devices (in particular of Apple's iPad), the project aims to provide school students who have no particular background in music, with access to the creative and well-being benefits of collaborative and personally expressive music making. This project takes a step forward from our previous network jamming research by using Apple's GarageBand [7] software on the iPad rather than the our own jam2jam software on laptop and desktop computers. jam2jam was specifically written for our previous research on how technologies afford meaningful engagement with music. It was used in this

A.R. Brown (✉) • A. Hansen
Queensland Conservatorium, Griffith University, Brisbane, QLD, Australia
e-mail: andrew.r.brown@griffith.edu.au; amber.hansen@griffith.edu.au

D. Stewart • A. Stewart
School of Public Health, Griffith University, Brisbane, QLD, Australia
e-mail: donald.stewart@griffith.edu.au; alanna.stewart@griffith.edu.au

© Springer International Publishing Switzerland 2014
D. Keller et al. (eds.), *Ubiquitous Music*, Computational Music Science,
DOI 10.1007/978-3-319-11152-0_4

capacity between 2002 and 2012. The main software features of jam2jam that support accessibility and engagement are (1) the use of generative music processes to enable participation by inexperienced musicians; (2) the ability for systems to be synchronised over a network facilitating coordination amongst users, either locally or at a distance; and (3) the ability to record music-making activities and export these for sharing. These features are now present in GarageBand for iPad (and an increasing number of other commercial software for mobile computing hardware). In our previous examination of developing resilience in school contexts, positive contributing factors included students developing a sense of autonomy and feelings of connectedness with peers and adults. We suggest that the scaffolding effect of generative music process can assist in promoting a sense of creative autonomy in inexperienced musicians and that the collaborative aspects of group music making can strengthen feelings of connectedness amongst peers. An aim of this project is to show how the principles of education and health promotion developed in our previous research can transfer to the use of ubiquitous computing systems.

4.1.1 Brief Description of the Project

This project focuses on building and supporting young people's engagement and connectedness with their creative selves and to help build resilience through musical collaboration and success. Working with a school of Indigenous Australian students (the Murri School) based in Brisbane, Australia, we have provided opportunities for musical expression using music technology through the school curriculum.

The project engages Indigenous Australian students using a digital audio production system that allows them to use their personal, social and cultural identities to make meaningful creative endeavours. The project trials newly emerging technology using iPads and the GarageBand software that has generative music and synchronisation capabilities, to explore the development of self-confidence and self-esteem.

The approach involves the trialling of weekly music-based activities in several classes over two terms (20 weeks). The activities are designed to offer opportunities for students to achieve creative educational goals, to engage them in expressive music making, to develop self-esteem and to develop creative collaborations with peers.

The project aims to provide evidence of a positive model for engaging school students in an interactive music-based education programme and for building confidence and resilience. The objectives of the project are to:

- Trial and evaluate new generative music technology to explore improvements in engagement and connectedness between students and the education system.
- Build resilience and raise the levels of educational achievement and aspiration of Indigenous students.
- Identify positive models of music education and health promotion.

- Use music technologies to build a sense of belonging and connectedness within the school environment that is protective of mental and emotional well-being.

4.2 Accessibility via Mobile Technologies

A catalyst for this project is the availability of appropriate computing software and hardware for music making. Apple's iPad and GarageBand software have features that make the activities of this project much more accessible than they have previously been. The iPad's small size and touch interface make it easy for students to handle and easy for schools to accommodate. The GarageBand software utilises "smart instruments" and "Apple loops" that simplify music production and recording and editing tools that allow personalisation and customisation. The smart instruments provided a constrained performance environment that minimises "mistakes" and can be used in music education in a similar way that restricted acoustic instruments (such as small xylophones) have been employed in the past. The music clips (Apple loops) allow for a constructor-set approach to music making where students can combine these building blocks without needing (yet) the facility to make the clips from scratch. GarageBand's audio recording capabilities enable students to capture and edit their own voice and sounds for use in their productions. The iPads and GarageBand combination support collaboration by allowing students, each with an iPad, to synchronise their music making over a local network. This activity, which we have previously called network jamming, facilitates groups of students to perform together. Finally, the ability of the software to record the music they compose and export files for later review and distribution means that students' works can be available for reflection and or sharing with the wider community.

4.3 Meaningful Engagement

The theory of meaningful engagement was developed by Andrew R. Brown and Steve Dillon [4] and has underscored the development of network jamming research more broadly. It involves two dimensions. Musical engagement includes various creative behaviours or ways of being involved in music. The modes of engagement outlined in the theory cover a range of interactions from listening and appreciating, to creating, performing and leading. The theory suggests that meaning can arise from engagements with music in three contexts; personal, social and cultural. That is, music can be personally satisfying, it can lead to positive social relationships, and it can provide a sense of cultural or community identity. Below is a summary of the modes of engagement and contexts for meaning.

	Appreciating	Evaluating	Directing	Exploring	Embodying
Personal	Listen, Read, Watch	Analyze, Select	Compose, Produce	Improvise, Experiment, Investigate	Practice, Play
Social	Share, Recommend	Discuss, Comment	Conduct, Lead, Instruct	Jam	Rehearse, Record
Cultural	Attend, Patronize	Curate, Review	Promote, Manage	Research, Publish	Perform

Fig. 4.1 The meaningful engagement matrix with exemplar musical activities

Modes of Creative Engagement

- Appreciating—paying careful attention to creative works and their representations
- Evaluating—judging aesthetic value and cultural appropriateness
- Directing—leading creative making activities
- Exploring—searching through artistic possibilities
- Embodying—being engrossed in fluent creative expression

Contexts of Creative Meaning

- Personal—intrinsically enjoying the activity
- Social—developing relationships with others
- Cultural—feeling that actions are valued by the community

The two aspects of meaningful engagement can be depicted as the axes of a matrix, as shown in Fig. 4.1.

The meaningful engagement matrix (MEM) is a framework for describing creative experiences and evaluating creative resources, plans or activities, for example, assessing a community or educational workshop, reviewing the comprehensiveness of an arts curriculum or lesson plan and evaluating the affordances of a software application for creating media content. While this matrix was developed for musical activities, it can be applied to other pursuits, especially in the arts.

Artistic experiences become meaningful when they resonate with us and are satisfying. The MEM has been developed to assist inquiry into our creative activities and relationships. A full creative life, the theory suggests, involves experiences across all cells of the matrix. Therefore, this framework can be useful when auditing the range of experiences afforded by any particular activity, programme or resource or across a set/series of these. It is in the assessment of the whole-of-programme view of this project that the MEM provides its greatest utility.

4.4 Resilience

Resilience refers to the capacities within a person that promote positive outcomes such as mental health and well-being and provide protection from factors that might otherwise place that person at increased developmental, social and/or health risk [6, 11]. Factors that contribute to resilience include personal coping skills and strategies for dealing with adversity such as problem-solving, cognitive and emotional skills, communication skills and help-seeking behaviours [6].

There is an abundance of research that highlights the importance of the social environment or social relationships for fostering resilience [8, 11]. Social cohesion or connectedness refers to broader features of communities and populations and is characterised by strong social bonds with high levels of interpersonal trust and norms of reciprocity, otherwise known as social capital [12]. This network of rich social relationships and strong connections promote a sense of belonging and community connectedness which, in turn, impacts on an individual's mental health and overall well-being [2]. Social capital, spirituality, family support and a strong sense of cultural identity are key protective factors for Indigenous people (and children) [9].

Schools that aim to strengthen their capacity as healthy settings for living, learning, working and playing, and are underpinned by inclusive participatory approaches to decision-making and action, can help to build resilience [11]. Connectedness in the school setting has been shown to be a protective factor of adolescent health risk behaviours related to emotional health, violence, substance use and sexuality. Creative activities, especially collaborative ones such as music making, share many of the characteristics that have been shown to promote resilience. This project seeks to take advantage of these connections.

4.5 Collaboration and Sustainability

With relevant support from the Murri School community, the project offered the opportunity to develop a creative and sustainable programme for young people, in this case young Indigenous Australians, to engage in collaborative music-making activities using interactive music technologies. Music technology is appropriate for this project because it is familiar to young people and because of our expertise in the use of generative systems in collaborative music making.

A number of creative projects use music jamming as a means of improving creativity, social justice and well-being [1]. The GarageBand software for the iPad supports collaborative audio production through local synchronisation via Bluetooth and through file and audio material export and import. When used as a musical instrument and compositional platform, this software enables students to build on basic skills of exploration and improvisation and encourages engagement. These technologies are also easy for staff to learn and use, and this, it is hoped, will increase the likelihood that the network jamming activities will continue in the school beyond the life of this project. A number of strategies were used to facilitate the sustainability of the activities. These include:

- Involvement of school administration and teaching staff in the planning and execution of the activities.
- Integration of the music activities into the broader curriculum.
- Sharing of the musical outcomes amongst the school community.
- Regular reporting on progress with the school administration.
- Provision to leave the equipment used for the project with the school.

4.6 Case Study: iPads and Music at the Murri School

The goal of the project was to examine how music technology can work to improve Indigenous health and well-being by creating a sustainable programme for indigenous youth to engage in collaborative music-making activities using interactive music technology.

The project integrated music activities using the iPad into the normal school curriculum and involved relevant teachers. It used standard classroom procedures and resources, but the project provided a facilitator proficient in the technologies and familiar with theories and objectives of the project. The project involved a weekly session with each class facilitated by a member of the project team and the class teacher (Fig. 4.2).

Prior to commencing, teachers and students were provided with information about the project, and teachers were consulted about how the music-based activities might integrate with existing curriculum objectives. Many teachers chose to incorporate creative writing tasks as the basis for song writing and rapping. The project used a whole-school approach, and classes were chosen from across the full age range of the school for participation. Students and teachers were not screened for

Fig. 4.2 Images from the project school

Table 4.1 Participating groups and activities

Year level	Approx. age	Activity objective
2/3	7/8	Students to write and record a short four-line rap about the good qualities they see in themselves
4	9	Students to record a creative interpretation of their sonic personal profiles utilising sounds and music to express their personalities
8	13	Students to write and record a sonic poem using text and music describing themselves and their hopes, expectations and dreams

musical background nor on any measure of resilience as we were keen to investigate the versatility and flexibility of this approach across the school community.

After consultation with staff, three grade levels were selected to participate in the project. The year levels and project summaries for these classes are summarised in Table 4.1.

4.6.1 Designing Music-Based Activities

Prior to facilitating the intervention with the students at the Murri School, a series of generic activities were designed with a view to facilitating creative participation in a way that adheres to the philosophy interwoven in the aforementioned MEM framework. These were used as a resource to stimulate activity design and lesson planning during the project. The key objectives of the music-based activities designed for this project were to (1) enable the students to engage in diverse music-making opportunities that utilise music technology in a meaningful way and (2) to enable the participants to have the opportunity to engage in creative experiences that assist in positively strengthening their sense of well-being and resilience.

Based on these resources, activities for each year level were collaboratively developed by the researchers and participating class teachers, keeping the MEM in mind throughout this process. Each teacher chose to utilise an age/ability-appropriate literacy basis for their class project in order to facilitate the opportunity for students to individually and collectively express themselves and their interests in a personal and creative manner.

The objective for Term 1 was to enable students of each participating group to develop and record their own composition using GarageBand on the iPads. The timeline below outlines the context of each weekly session dedicated to the project, allowing for students of each group to spend time experimenting, jamming, practicing playing and recording instrumental sounds and external audio and for recording the final product. The objective for Term 2 was for students to develop and refine their work into a form ready for a "signature" event—a public performance at the school assembly.

Table 4.2 lists the mode and context of activities designed to achieve the key objectives of this project. Each cell corresponds to specific mode and context combination within the MEM.

4.6.2 *Measuring Resilience and Engagement*

Evaluation of this project relied on a mixed methods research design combining quantitative and qualitative methods of data collection, analysis and inference in order to investigate both the processes developed through the life of the project as well as the impact of the project over time. Students were asked to complete a modified version of a pre-existing resilience questionnaire that has high levels of reliability and validity [5]. Key informant interviews with staff were conducted and subject to an ongoing thematic analysis. An introductory school consultation session was attended by nine staff members at the outset of the project. All were supportive and identified ways that they could integrate the project into their curriculum. Due to timetabling constraints, only three of these staff and their classes participated in the project. Thirty-four students participated in the project across three grade levels: years 2/3 (14 students), year 4 (12 students) and year 8 (eight students in the English stream). Activities included developing a rap, recording a personal sonic profile and writing and recording a bio-poem. Observations of class sessions were recorded in a journal by a member of the research team. In addition, files of work completed on the iPad were regularly saved allowing for analysis of the steps taken during the creative process.

4.7 Survey Results Summary

The first stage of data collection provided a baseline, and descriptive statistics show some differences between the younger students in grades 2/3 and 4 and their fellow students in grade 8. We have not completed tests of statistical significance as the sample is small. We provide, below, a selection of the results and findings. First we provide a summary with some data from the first resilience survey, conducted prior to the students participating in the activity, to give a sense of the student's attitudes and expectations from the project. Over 75 % of the total student sample thought that being involved in the project would be fun, and most (younger students) were excited at the prospect. The creative levels and aspirations of the students were uniformly high, and almost all indicated that they enjoyed going to music performances. However, compared to the grades 2/3 and four students who relished

Table 4.2 Project activities across the meaningful engagement matrix

		Attending	Evaluating	Exploring	Directing	Embodying
		Listening/observing	Reflecting/analyzing	Experimenting/improvising/conscious	Decision making/instructing	Playing/performing/establishing habits
Personal (of the self)	Objective	Independently listen, read and observe in order to become aware of relevant knowledge	Independently reflect and analyse personal practice as a means of facilitating continued learning	Independently explore and experiment with relevant artefacts and processes	Engage in technical activities that lead to creating a musical artefact	Independent practice/playing
	Activity	Introduction to network jamming. Demonstration of available interactive music hardware and software	Record/journal learning and practical experiences. Music analysis to enable the development of aural skills	Independently explore and experiment with sounds and functions of network jamming devices. Building knowledge	Setting up a jam session. Composing a song	Guided and independent play/practice of network jamming devises and processes to build-ing skills
Social (collaborative)	Objective	Share work and progress with peers	Reflect upon learning and practical experiences with peers as part of group discussions	Extend learning through collaborative experimentation	Take on a leadership role within a group activity	Rehearse and record with a group

(continued)

Table 4.2 (continued)

		Attending	Evaluating	Exploring	Directing	Embodying
		Listening/observing	Reflecting/analyzing	Experimenting/improvising/conscious	Decision making/instructing	Playing/performing/establishing habits
	Activity	Workshop presentations of individuals and collaborative engagement and progress with network jamming	Group discussion	Engage in a group (networked) jam session	Lead and conduct a jam session. Group composition	Time to play/practice with network jamming devices and processes collaboratively.
Cultural (connection with external)	Objective	Observe relevant activity as performed in a public context	Extend and connect reflective practice to include a wider cultural participation and dialogue	Examine/research relevant practice in a wider cultural context	Support and promote a musical artefact for public distribution	Participate in a group public performance
	Activity	Attending/observing a performance that utilises network jamming as a key composition/performance process	Develop a creative project for public presentation. Create a blog/website as a reference for music work	Investigating network jamming in diverse cultural contexts. Explore other commercial music apps	Create and promote a CD/DVD showcasing creative progress	Perform a group "jam" or composition to an audience

the creative opportunities of the project, a substantially lower percentage of the grade 8 sample felt confident and supportive of the activity and their creative role.

1. With regard to their confidence with and support structure for creative activities:

   - Over 85 % of all students like making things that are creative and different.
   - Students felt variously confident with their own creative ability and ideas. (71 % of grade 2/3, over 90 % of grade 4 students, grade 8 = 63 %).
   - Most students have family/elders that they can go to for help (grade 2/3 = 79 %, grade 4 = 90 %, grade 8 = 75 %). The students' attitudes toward peer collaboration varied between the younger and older students.

2. The following data reflect these attitudes to working with classmates:

   - Students like to share their creative ideas with their classmates (grade 2/3 = 78 %, grade 4 = 90 %, grade 8 = 37 %).
   - Students enjoy hearing about their classmates' creative ideas (grade 2/3 82 %, grade 4 = 85 %, grade 8 = 63 %).
   - Students thought that being a part of the project would help them have more friends (grade 2/3 = 75 %, grade 4 = 75 %, grade 8 = 12 %). As with attitudes to collaboration, the students' sense of self-confidence in public music making also reduced with age.

3. In relation to producing a performance or recorded outcome:

   - Students thought that they could put together a performance or recording that would be enjoyed by others (grade 2/3 = 86 %, grade 4 = 66 %, grade 8 = 12 %).
   - Students felt that people would come to watch their performance or record launch (grade 2/3 = 90 %, grade 4 = 75 %, grade 8 = 25 %).

A clear trend in this data is the difference in reported self-confidence, in music at least, between the younger (7–9-year-old) and older (13-year-old) students. This is consistent with much more extensive research that shows a dip is self-confidence in adolescents [10]. As a result of this, and supported by informal feedback from the grade 8 teacher, we adopted a different strategy for the older group. Activities for this class focused more on personal meaning than on social or cultural meaning, and we tried to minimise potentially embarrassing public presentations of the music. As well, work for older students has a greater individual focus whereas activities for younger students are heavily biased toward group work and include class and public presentation of outcomes in the form of recorded media and live performance. What is interesting to note is that the accessibility features of the music technologies employed are equally applicable for both groups and approaches.

4.8 Qualitative Results Summary

Qualitative data included interviews conducted with teachers, notes maintained by research team members and samples of media produced by participants. The research team utilised the MEM to record the frequency and intensity of meaningful engagements they observed in students participating in the project. Video footage and photography were also being used to provide further documentation of project implementation activities and for review and analysis.

4.8.1 Pre-intervention Results

Staff members recorded their initial plans for implementing the project within their classrooms for Term 1 and Term 2, 2013. Eight out of the nine staff members participated in this component of the staff session. Participant responses to what they hoped to achieve by being involved in the project include:

- For the students to record stories created for English unit. The story can be edited and compiled onto a CD. Hopefully children will gain confidence in speaking and sharing their stories/ideas.
- "I would like to see students engage with iPad technology to enhance and extend learning already happening in subjects."
- Improve teacher and student confidence and participation with technology; having children work together cooperatively; tap into children's different learning styles, i.e. rap songs to learn spellings; student enjoyment.
- To use the jamming as a learning/teaching tool in classroom—to integrate curriculum to make learning fun.
- "To learn myself and get children involved in expressing themselves orally and musically."
- To record for a performance and to make learning fun and for students to use an iPad.
- Enhancement of student work (oral and written)—familiarity with technology.
- Increase iPad literacy, learn with students how to use this tool for work.

4.9 Post-intervention Results

4.9.1 Classroom Management

In terms of general process, the participating classroom teachers had differing opinions regarding how manageable it is to have a class of students work with the iPads for engaging in learning and collaborative work. Two of the teachers felt that this was a manageable task, whereas one of the teachers (grade 4) felt that this process of learning would work best in smaller groups as children may have difficulty listening to instructions and paying attention in a larger group. Some of the challenges in participating in the project include student's inability to share iPads. They preferred to work on their own. Another challenge lay in having a consistent and clear idea of the long-term goal and clarifying goals for students to be achieved at the end of each session.

One teacher felt that at times it seemed that the students were "all over the place". This was due to the students showcasing their ability to "jam" on the iPads using different musical instruments available on GarageBand. Jamming with colleagues allows for creative expression that relies on self-expression. The grade 4 teacher felt that not being present regularly and not understanding how to use the iPads and remembering it were challenges. Also, keeping all the students on task when the whole class was involved was a challenge. She felt that keeping the iPad project in a small group environment might assist in overcoming some of these challenges. However, in terms of how satisfied the teachers were with the way the project had been implemented in their class, there was general consensus that they felt that the project went well and that the students looked forward to the sessions on the iPad.

4.9.2 Student Engagement

Observational notes and a review of the media outcomes reveal that students engaged in a variety of ways consistent with the spread of modes in the MEM. However, as might be expected, the depth of engagement varied between individuals and within individuals over time. For example, some of the year 8 participants took an extended time to develop material for their sonic poems but regained enthusiasm when this material was combined with semiautomated backing tracks for the recorded outcome at the end of the first stage. By way of contrast, the younger classes started with a great deal of enthusiasm and were only moderately interested in the recorded outcome but were very enthusiastic in the second half of the project that focused on a performance outcome. The grade 2 teacher was really impressed that his more challenging students, who rarely engaged in classroom activities, were able to participate confidently in the project. Those that had difficulty with directing their attention to one specific task for a period of time were able to participate in the iPad sessions for the course of the weekly schedule. The grade 8 teacher felt that he underestimated the students' reluctance to share their work. He felt that his lack of knowledge of technology/iPads required increased reliance on the project facilitators. He acknowledged that the students had a product at the end of the project but considered that the iPads could have been better used.

4.9.3   Teacher Engagement

The grade 4 teacher felt that there were components of the programme that she liked and some parts of the programme she did not find helpful in making the project run smoothly. Teacher ownership is a critical success factor for the sustainability of the project. She felt that because she wasn't there most of the time for the weekly iPad sessions, she found it difficult to gauge how effectively the implementation was going. She indicated that there were times when it was confusing what the object of the lesson was. This reinforced the importance of working with the teachers to develop an action plan for their students and take a leadership role in achieving their goals and objectives. The participating teachers relied on the project facilitator to set weekly plans for the students, offering limited guidance and support. Behaviour management was a challenge each week for the facilitators. Often teacher aides were the only other adults present to provide additional supervision for the children, and at times sessions were taken up with disciplining students.

4.9.4   The "Signature" Event

The grade 4 teacher stated that he enjoyed watching those children who performed on assembly and that, in the end, the project performance sounded good. He stated that some of the students are normally really shy and would never get up on their own. But, because they were in a group and focusing on the iPad they coped. All teachers stated that they were happy with what their class had achieved by participating in the project. The grade 8 teacher stated that hopefully they will have greater confidence to use technology to support English language leaning with music-based activities.

4.10   Findings

This project applied the practices of network jamming and the use of the MEM to the use, for the first time by these researchers, of commercial software in the form of Apple's GarageBand for the iPad in a school setting. There was a particular focus on the use of musical activities to support student resilience. Overall the application of these mobile technologies was successful both practically in providing ease of access and in allowing for the range of meaningful engagements outlined in the MEM.

The technologies supported private, social and public music-making practices. These were applied to varying degrees as seemed appropriate to student's levels of confidence and capability. For some students, their focus was on private creative expression, others worked collaboratively in networked classroom situations, and some students participated in a public concert of original works using the technology.

Observational analysis showed that students were able to engage in activities that involved various modes of engagement, from listening, to exploring, creating and performing. The cross-curricular integration of musical tasks, especially to support language development and expression, was well received by students and teachers. While there was no formal measure of an increase in resilience, the adaptability of the pedagogical approach to suite a wide range of personality types and degrees of confidence was noted by teachers and facilitators.

All teachers considered that their involvement in the project has made a difference to the way they have looked at teaching. The grade 4 teacher stated that it gave her another avenue through which to teach. Technology is the focus of our learning now, she said. The grade 8 teacher stated that it has highlighted a need to use technology in the class. Students have access to it outside of school they use it all the time—it is a tool he feels he needs to tap into for learning. All participating teachers have plans to continue to use this form of learning for future teaching.

The project indicates that it is beneficial to have a structure around using the iPads in class. To start with structure was thought to be important, i.e. weekly plan/within a subject. The grade 8 teacher felt that freedom to be creative can flow on from this.

Indicators are that the project had had a positive impact on the students. The grade 8 teacher stated that the students looked forward to "Friday" sessions. He stated that although they were shy, he believes that they were secretly proud of what they did. The grade 4 teacher said that they loved it and looked forward to it. She also said that she could use the iPads as a reward for good behaviour.

All teachers stated that they would recommend using the iPads as an approach to learning to other teachers. The grades 2 and 8 teachers felt confident in sharing this approach to learning with colleagues. All teachers felt that they would have liked more professional development on using the iPads.

4.11   Lessons Learnt

This project aimed to examine how music technology can work to improve student health and well-being. The project aimed to offer the opportunity to develop a creative and sustainable programme for young Indigenous Australians to engage in collaborative music-making activities using interactive music technologies. The following lessons have been learnt from this pilot project:

- An in-class project of this nature requires relevant support from the whole Murri School (Indigenous) community.
- A planned period of in-service training and support with the teachers would help to ensure that the project is introduced with confidence and becomes sustainable beyond the life of the project.

- Small group work with all students accessing the technology would ensure better student engagement.
- A clear link between curriculum frameworks and the use of iPad technology helped to engage the project within the School's learning community.
- Students enjoy and engage with mobile media technologies within the classroom and can develop meaningful experiences through personal and social creative expression.
- The MEM provides a strong theoretical framework for planning a school-based creative project. Further data analysis might also confirm its utility for analysis of student behaviour and experience.
- Additional research is needed to confirm the reliability and validity of the questionnaire with consideration given to a range of instrument structures to allow for widely varying age/developmental conditions.
- This project provided a constructive and stimulating experience for many young people, even those who find group work difficult and have communication difficulties.
- Public performance of creative, music-based projects provide important opportunities to enhance self-esteem and promote creative partnerships.

Conclusion

In this chapter we have described our use of mobile technologies and software to make music-based activities accessible to young people in a way that promotes meaningful engagement and resilience. The project was based in the Murri School in Brisbane, Australia, that is dedicated to the education of Indigenous Australians. The project involved weekly activities with three classes from that school over 20 weeks with students ranging from ages 7–13. The design of project activities was informed by theories of meaningful engagement and resilience but were guided by the advice of class teachers and student survey responses to ensure appropriateness to the local context. Data indicate that staff and students are enthusiastic about using the tablet computers and music apps and that their ease of use is making previously unimagined music production activities accessible. Consistent with other studies, our data shows a dip in the creative self-confidence of students in their early teens (compared to younger students). This has been accommodated for by shifting the emphasis for those students toward individual and personal expression and away from collaborative and public activities. The portability of the iPad hardware has assisted with the integration of the devices into the school environment, and their multipurpose nature makes for fluid shifts between music and other curricular tasks (such as creative writing). The GarageBand software has facilitated rich music production outcomes, although the devices alone provided limited audio recording and playback

(continued)

quality. We plan to address this in any future projects through more extensive use of external microphones and headphones. Indications are that the students can be keenly engaged in network jamming activities but require an ongoing facilitator support to maximise creative outcomes. The features of the music-based activities with ubiquitous technologies align well with characteristics that promote resilience, including personal autonomy and connectedness with peers and adults, and we remain optimistic that evidence of a positive effect on student resilience from the project can be achieved.

Acknowledgements We would like to thank the staff and students of the Murri School for their participation in this project. The project was supported by the Queensland Centre for Social Science Innovation. We'd like to acknowledge the vision and enthusiasm of the late Steve Dillon who instigated this project, but, unfortunately, was not able to be part of its realisation.

References

1. Adkins, B., Bartleet, B.L., Brown, A.R., Foster, A., Hirche, K., Procopis, B., Ruthmann, A., Sunderland, N.: Music as a tool for social transformation: a dedication to the life and work of Steve Dillon (20 March 1953–1 April 2012). Int. J. Commun. Music 5(2), 189–205 (2012)
2. Australian Institute of Health and Welfare: A Picture of Australia? Children: Health and Wellbeing of Indigenous Children. AIHW, Canberra (2009). http://www.aihw.gov.au
3. Brown, A.R., Dillon, S.C.: Networked improvisational musical environments: learning through online collaborative music making. In: Finney, J., Burnard, P., Brindley, S., Adams, A. (eds.) Music Education with Digital Technology, pp. 95–106. Bloomsbury Publishing, London (2007). ISBN 9781441186539
4. Brown, A.R., Dillon, S.: Meaningful engagement with music composition. In: Collins, D. (ed.) The Act of Musical Composition: Studies in the Creative Process, pp. 79–110. Ashgate, Surrey (2012)
5. California Department of Education: California Healthy Kids Survey (2004)
6. Fraser, M.: Risk and Resilience in Childhood. NASW Press, Washington (1997)
7. GarageBand: iOS Music Software (2014). https://www.apple.com/ios/garageband/
8. Maggi, S., Irwin, L.G., Siddiqi, A., Poureslami, I., Hertzman, E., Hertzman, C.: Knowledge Network for Early Child Development. World Health Organisation, British Columbia (2005)
9. Malin, M.: Is Schooling Good for Aboriginal Children's Health? The Cooperative Research Centre for Aboriginal and Tropical Health, Northern Territory University (2003)
10. Orenstein, P.: Schoolgirls: Young Women, Self Esteem, and the Confidence Gap. Anchor Books, New York (1994)
11. Rowe, F., Stewart, D.: Promoting connectedness through whole-school approaches: a qualitative study. Health Educ. 109(5), 396–413 (2009)
12. Siddiqi, A., Irwin, L.G., Hertzman, C.: Total environment assessment model for early child development (2007). www.who.int/social_determinants/.../ecd_kn_evidence_report_2007.pdf
13. Stewart, D.E., Sun, J., Patterson, C.M., Lemerle, K.A., Hardie, M.W.: Promoting and building resilience in primary school communities: evidence from a comprehensive 'health promoting school' approach. Int. J. Ment. Health Promot. 6(3), 26–31 (2004). http://eprints.qut.edu.au/1281/

Chapter 5
Analogue Audio Recording Using Remote Servers

Lucas Fialho Zawacki and Marcelo de Oliveira Johann

Abstract This work proposes a system for remote recording using real analogue instruments and equipment over the Internet and analyses several possibilities for its implementation and usage. In the first example, which has a running prototype, a file with a MIDI performance is submitted to the service's site, being played with an actual analogue synthesiser, and a high-quality recording is made available back to the user. The proposed service model is characterised by batched access, processing automatically a queue of submitted jobs in such a way as to maximise the hardware utilisation and therefore allowing very low accessing costs for the end users. A significant contribution of this work is to show that a much higher efficiency is attainable with the proposed service model, what is not possible with both local or remote real-time systems. We proceed to analyse the possibility of implementing a variety of other analogue and acoustical processes for music production as remote audio servers, providing quality and cost-effective remote access from anywhere to anyone. The chapter discusses motivations and context in further detail, addresses major compromises, the required technology, and describes the implemented prototype. The ultimate goal is to allow musicians to compose, record and mix their music from simple tools in their computers but having the opportunity to access also very high-quality analogue and real instruments for final sound rendering. In this text, both digitally controlled and old analogue synthesisers are considered, as well as adapted and custom instruments, acoustical musical instruments of many sorts, effect processing units, analogue mixing consoles and a few others. A service like this can also make expensive or rare machines available to a wide range of users. Each different class of tools has its own context, requirements and challenges, for which possible solutions are discussed in this text, that also highlights the many commercial and creative possibilities of such an ubiquitous approach to analogue audio.

L.F. Zawacki • M. de Oliveira Johann (✉)
UFRGS, Porto Alegre, RS, Brazil
e-mail: lfzawacki@gmail.com; johann@inf.ufrgs.br

© Springer International Publishing Switzerland 2014
D. Keller et al. (eds.), *Ubiquitous Music*, Computational Music Science,
DOI 10.1007/978-3-319-11152-0_5

83

5.1 Introduction

Throughout history, humankind has produced a large variety of musical instruments, many of them tied to different styles, cultures or historical moments. Each instrument has its unique sound and aesthetics that can be explored in the appropriate context. Modern music seeks to embrace this entire diversity, oftentimes driven by the movies and gaming industries that cover a diverse cultural background in their works. The advent of computers and digital audio created new forms of sound composition and new instruments, which became part of our culture and additional tools at our disposal. The demand for different sounds can be supplied in many ways. On the one hand, real original instruments can be used, but they can be sometimes hard to access or expensive and the services of specialised studios may be necessary. On the other hand, computers can provide virtual instruments, and this has been a very popular option with software emulations in place of real instruments.

Both options have their advantages. Real instruments generally have the best sound quality and smooth control, while virtual instruments have lower cost while maintaining a relatively good quality. The big market for virtual instruments, mostly in the form of plug-ins for digital audio workstations (from now on called DAWs) shows that the search for different instruments and aesthetics is ever more common in all kinds of productions. But users are always looking for the best of both worlds, flexibility and low cost of virtual instruments and a quality closer to the real ones. Music production is nowadays a much more common activity and is done by professionals and amateurs alike, as can be evidenced by the growth of independent artists in online communities [27]. We can say that the need for quality instruments and audio processes is shifting from niche to ubiquitous. So, we come to the difficult question of how to improve the quality of Ubiquitous Music [20] not being constrained to the limitations of the digital plug-ins. We do not want to avoid or dismiss virtual instruments completely either but understand that both options should be available when needed.

To do so we propose a system that offers remote access to real instruments as well as to analogue equipment and allows the recording of the resulting sound. There are many works that deal with the problem of live music using the web, real-time access to remote instruments and all associated issues as, for example, [19] or [1]. This work does not try to explore or tackle these problems but goes in another direction. Instead of real-time access and streaming, we propose a system that sacrifices these characteristics to provide better flexibility and, low cost and still maintain the highest sound quality. This system also uses online web access but processes user requisitions in batch format and serves them by scheduling the jobs.

In such proposal, the system is not locked to a given user in a time frame, and there is no problem of simultaneous access, neither latency nor streaming issues. The system behaves as a contract in which users send their job requests, and after some time, they receive back the resulting service in the form of a high-quality recording of a real analogue or acoustical instrument. In this way, it can enrich the available toolset of professional and novice musicians, allowing ubiquitous access to real instruments.

5.1.1 Proposed Architecture and Access Model

A key aspect in the proposed system lies in the batched access model. It works like an analogue server to which the user submits a job and retrieves the expected result after some short time. This is a different scenario compared to real-time access, which requires tight timing considerations between user and server, with latency and throughput issues, as addressed in many works on network online music [14, 19]. The main purpose for the batch access is to increase system usage. Real-time access to an analogue synthesiser implies that whenever the system is being used or available to a given user, it cannot be accessed by others. In other words, we need to serialise the users, something that cannot be implemented efficiently. The only way to do that is to reserve big time frames (like half-hours) for each user, resulting in poor system usage. In the proposed system, instead of being able to play with the synthesiser in real time, like in [15], the user is expected to prepare a job and submit it to a queue, where it will be processed shortly. To our knowledge, this is the first work that addresses batched access to analogue synthesisers.

The architecture of such a system can be seen in Fig. 5.1. The system must be composed of a front-end, a hardware infrastructure and a software processing system that implements the desired functions. Starting with the hardware, at least the real synthesiser, a computer, an audio interface and a MIDI interface are needed.

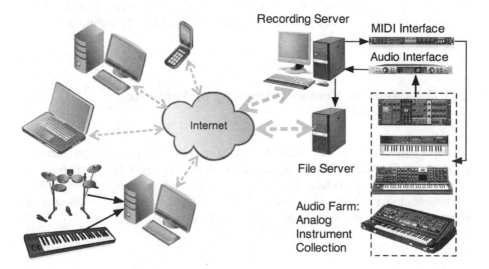

Fig. 5.1 System architecture and access model

The job can be described by a MIDI performance or employ another music notation format [12]. But MIDI remains the most popular and by far the most supported in musical interfaces that are available to end users. Therefore, it is almost mandatory to support it, as it is done in the first prototype described ahead. The software system must be able to play the performance through the MIDI interface and record back the resulting audio with an audio interface. The software system has at least two other attributions, which are to implement the queue, dispatching each job with the proper synchronisation, and to make a set of pre-processing tasks. Among those, it is important to check the user-submitted MIDI file for errors, bounds and filter out undesired messages. The last component is the user front-end. It must allow the user to submit its performance, select a set of parameters and get access to the resulting recording. The MIDI submission can be implemented from inside a plug-in running in a given DAW or with a standard web page, as implemented in our prototypes. This requires the user to export each one of its MIDI performances from its DAW to a MIDI file on its computer's file system, from where it can be further uploaded to the service.

A running prototype of such an analogue server that can be used with the web is briefly described in the next section. Based on the observations with that implementation, a set of wider possibilities was realised. We then proceed to provide a prospective study of what kind of instruments and analogue audio processes could also be accessed remotely, under which circumstances, and what are their main requirements. Considering the entire project, the main contribution of this work lies in the concept of analogue audio server, or analogue audio farm, associated with the usage model, through which performance jobs are submitted to be rendered in a batched (not real-time) system, for high-quality rendering. In particular, we find that there are many other audio processes besides analogue synthesisers that can benefit from this model. We show that there is already available technology to implement many of these servers and highlight what are the main challenges involved to seamlessly doing it.

5.2 A System for Remote Recording of Analogue Synthesisers

In [29], a system for remotely accessing analogue synthesisers as analogue servers, using the World Wide Web, is proposed, and a working prototype is described. In this section we provide a detailed description starting from the basic idea, implementation, challenges and possible solutions, to the market opportunities, so that we can analyse and develop the concept further.

5.2.1 Motivation

The main motivation for this work was found in the importance that analogue synthesisers have for today's music production along with the increased prices of both old and newer such machines.

Analogue synthesisers started to be replaced with the advent of the digital technology in the 1980s. There are many reasons for this trend to have been so hardly pursued. Powerful analogue machines were big and expensive, their programming could not be easily stored, and they did not have a perfectly stable tuning, among others, all very important factors. They were also limited in their ability to reproduce complex initial transients that are characteristic of instruments like the piano.

However, today, more than three decades later, we must acknowledge the importance that analogue synthesisers still have. The synthesisers and computer software (plug-ins) available today are not only used to produce sounds that only digital technology can generate but also try to mimic the old technologies, acoustical instruments and the sounds of many analogue synthesisers. This is already a sign that they form a class by themselves, real musical instruments capable of characteristic performances and nuances. Actual analogue synthesisers are still heavily employed in modern pop music [21]. Descendants of the Moog synthesisers [17] are used to provide the deep basses in genres like rap, dance, and effects in electronic music, as well as solos in countless jazz and rock groups, just to mention a few examples. In summary, it is safe to say that while the digital revolution brought us incredible possibilities, better control, new sounds, everything widely accessible from simple consumer computers, the original analogue machines that are already part of the music aesthetics will continue to be considered important and irreplaceable in many aspects for years to come.

There is an accessibility issue, however. With few exceptions, analogue synthesisers, both new and old, are more expensive than before. Their construction must employ old techniques, components or practices that are not the mainstream today, while the original ones are getting older, so finding them in good shape is becoming harder. There are indeed collectors that have a good set of original machines, but just few people can access them, and new generations might not even get the chance to know how they sound, led to think that the plug-ins are accurate reproductions. There is an "analogue renaissance," currently happening, with many new and more popular products being released, specially by the Korg company, which sells pocket analogue synthesisers and faithful recreations of some famous models. It alleviates somehow the problem of price and availability, heating up the market. However, the cheaper models are subject to many technology limitations, like single-rail low power circuits, and do not compare to high-end models adequate for studio work. The movement indeed reinforces the need for accessing the class of more expensive machines.

So to address this need, we propose the compromise of remotely accessing the instruments in batched form. There is an intrinsic big advantage of the batched access strategy compared to the other options of real-time access or physical access (renting instruments or studio time). With batched access, the system usage can be kept high, avoiding waiting times, the limitation of arrangements for time slots, and preventing expensive equipments from being idle most of the time.

Put in other words, be it in a bedroom studio or in a very expensive facility, there are many expensive synthesisers (or other musical instruments) that are turned off, not being used, idle, in the majority of the time. That translates to large potential waste and also to high costs. When someone is renting a studio, he/she is paying for the contents (its capabilities, equipment), and most of the resources are not being used. Even the instruments that need to be used will be accessed sparsely, due to setup times, experimentation, trials and fails. We also need to take into account the constraints for studio time allocation. The user needs to find and to adapt himself to the available slots; he has to spend transit times and waiting time, so from his perspective, there is even more suboptimality. The only way to avoid those constraints is to acquire instruments and studio equipments. This is obviously expensive. While there are many artists that can afford to have their own high-quality equipment, it does not represent the typical user or musician. This is far from being popular, ubiquitous, far from encouraging widespread use of the instruments that humankind could develop throughout history.

In the proposed model, the equipment that is operated as an analogue audio server can be kept running with a high utilisation rate, provided that there is public interest and demand. In this way, it potentially alleviates the aforementioned problems. Notice, however, that we must not consider it as a pure replacement model. This is not being proposed as an alternative to people that can afford and effectively use instruments in an intensive way. As an example, a prominent keyboardist that studies everyday and gives frequent concerts with his band would not be expected to avoid having his own preferred synthesisers at home or in the studio. But a producer can perhaps access the sound of different synthesisers not available in his studio, for a given production or composition, from an online collection, that otherwise would be hard to find, or to pay for, if not operated remotely.

5.2.2   Working Prototype

The first prototype was implemented as a proof of concept for this service and is currently working with a small number of alpha users. The goal for this implementation was to check its viability using standard technologies and also test the access model and usage issues. The system offers access to a Studio Electronics ATC-1 with the Moog-type filter installed [6], a synthesiser that inherits most of its circuits from the Minimoog, and also to a Roland MKS-50, the module version of the original Alpha-Juno 2 polyphonic keyboard. The synthesisers are recorded using the Steinberg Cubase 6 Elements DAW and custom ADC hardware. The whole setup is running in an iMac connected to the synthesiser with a M-Audio MIDI interface.

The web interface is implemented using the Ruby on Rails framework and allows the users to upload MIDI files and create requests for recordings of these files, which will be entered into a queue and executed by the back-end server providing access to the synthesisers.

To actually execute each job, the back-end server implemented in Java takes the request from the queue and runs a custom AppleScript sequence. This sequence provides the automation of the DAW to create a new project, import the MIDI file, play it, record the audio and close the project after completion. The automation is possible using a combination of shortcut keystrokes and GUI operations, commanded with the AppleScript language. It is partially performed as if someone was using the computer, and for that, it includes a set of delays (0.5 or 1 s) so that the GUI system has time to create its elements before the next command is fired.

The interface implementation went through a few iterations and addresses many usability challenges faced by users. For example, some testing suggested that while MIDI is a universally used music format, the average user is not used to manipulating it directly and can get confused by concepts like tracks and channels. More often than not, the resulting recording would not meet the user expectations because of some parameter mistake or because the user uploaded the wrong file. To address that, the system presents some information about his uploaded MIDI file beforehand, like the name of each track, which MIDI channels it is using and how long the file takes to play in its entirety as shown in Fig. 5.2.

Before selecting which synthesiser and which programme change to use, the user can browse some pre-recorded samples of the given sound, and the same MIDI file can be used to generate different recordings using variations of synthesisers and/or programme changes. These recordings will all be listed under the same file in the user's account, as seen in Fig. 5.3, so that it is easy to experiment and find the right sound.

After the recording job is requested, the user is presented with a scheduled time when he/she will be able to download the file. The user is then notified via web interface when the recording is ready. The file is available both in WAV and FLAC formats for high-quality production and in MP3 and OGG using low bit and frequency rates only for the purpose of a web preview with a small file size as shown in Fig. 5.4.

5.2.3 Sound Selection and Preview

The major drawback of using a system that renders the sound out of real time is the lack of instant sound feedback, feedback which is both expected and sometimes absolutely required by the musicians so that they can produce the sound they want. Real-time interaction with the synthesiser is not possible. Another problem is the programmability of the synthesiser parameters. Even though MIDI is a standard way to communicate parameter changes to the hardware, the way synthesisers implement these changes varies considerably. A composer might want to have a more accurate

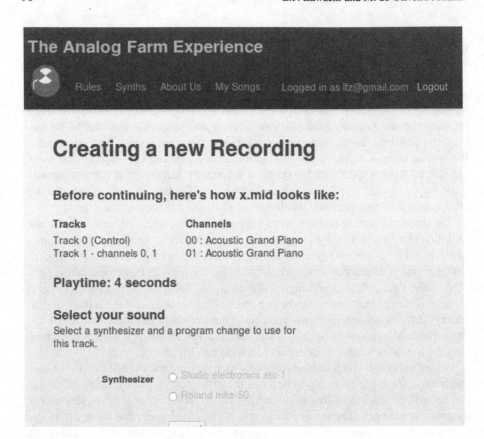

Fig. 5.2 Preview of the recording screen for one file

and predictable control of the programmable aspects of an instrument, and this is something such a system should be able to provide. There are a few options that can be used to work around both problems:

- Pre-recorded samples: There must be at least a standard sample for each sound, stored in a different server, so that the user can browse and listen to them to choose the one she needs. This is already implemented in the working system, but it must be improved to let the user get the desired results. In a simple example, the user should be able to preview not only the sound itself, but should also know the assignment from keys/notes to sound octaves.
- User feedback before recording: A short segment of the submitted job can be recorded and given to the user before processing it completely. In this way he can check the sound and parameters, as volume, base pitch and so on.

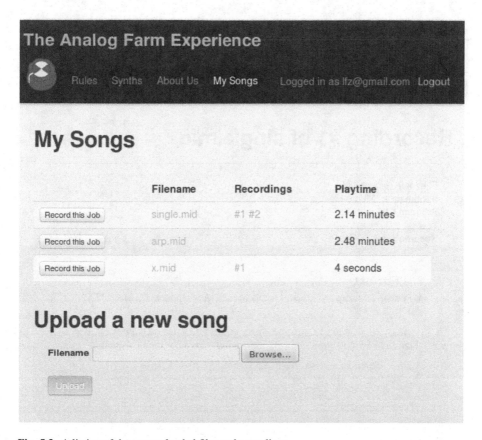

Fig. 5.3 A listing of the user uploaded files and recordings

- Virtual instrument for preview: Software plug-ins can be provided to mimic the sound of each programme in the actual synthesiser. With this option, the musician can prepare its production as if she were listening to a low-resolution copy of it, rendering the final high-quality option at the end. Considering that many companies today offer plug-ins that emulate classic analogue synthesisers, those plug-ins could be calibrated (or the system calibrated) so that the sounds and parameters closely match. A database of calibrated sounds can also be developed by the users, increasing its completeness with very little investment.

Fig. 5.4 Recording download screen

5.3 Usage and Operation Opportunities

In this section we explore a set of possible ways in which the system can be implemented and operated, together with business model opportunities. Let us start analysing typical usage cases and interpreting the possibilities that this service can bring to the user.

5.3.1 Typical Uses

A typical use would be to have a mixed MIDI/audio production in the user's DAW, with some tracks recorded with actual instruments and keyboards, and probably many of them with virtual synthesisers in the form of plug-ins. Some of these parts may need emphasis and call for a higher quality, like in a solo or bass line. The user can then browse a set of sound samples from the system, select one and submit its jobs to record one or a few tracks with real analogue synthesisers, increasing their quality in terms of timbre, articulation, smoothness presence, for example. Musicians already know that such sounds stand up in the mix [21], unlike their digital emulations. If used for a few tracks or shorter passages, even moderate costs would be perfectly acceptable, mostly because the user would pay only for the duration of the performances on those parts.

If the accessing costs are really low, as we estimate ahead, bedroom or garage musicians, as well as experimentalists can risk more trying to get different aesthetics or be surprised with unexpected sound effects. Among the experimentalists we can include people that sample and heavily process audio sources. This group is always in search of new sound sources that can be processed further, and they can access the system to get just a few seconds of some sounds generated by a particular analogue machine.

In a completely different environment, producers and session musicians have an enormous experience and usually know exactly what they need for that part being produced. They want total control and the ability to predict the resulting sound. As professionals, however, they are also used to know how things will sound at the end even when they are hearing only a preview. So it may be perfectly acceptable for them to work with software plug-in sounds during the major part of the production and then ask for a higher-quality rendering at the end, even for most of the tracks. This is exactly what happens in computer graphics for animation movies, where the animators work with sketches and render the high-quality images at the end on big computer farms. If this is an important production, they also should be able to afford submitting many times the same part, experimenting with slightly different sounds and tuning parameters so the final sound gets exactly the way it should be.

5.3.2   Aggregate Value and Market Span

Even for someone that has access to many interesting analogue synthesisers, recording them properly requires expensive analogue/digital converters (ADCs), preamps, wiring, a good Clock source and so on. When implementing a system that does recording exclusively in large scale, we can employ more advanced circuits specialised hardware, tune the recorder to the signal being recorded, offer a range of different sound flavours and employ many tricks that are challenging even in a top-level studio. This is possible with custom and expensive hardware that would be only affordable with dedicated high-usage systems, as detailed ahead. Based on the situations mentioned above, we speculate that the system can be used by a very broad audience, from bedroom studios to major labels, if properly implemented.

5.3.3   Relationship with the Traditional Options

A few questions arise as we consider it for commercial application. How about
the alternatives already used by the music industry today? Will a system like
this substitute them, decreasing their market? Should someone avoid buying an
analogue synthesiser and use the service instead? Should someone avoid buying
a virtual analogue plug-in? Is the system exclusively for people that do not have
analogue machines? The answer for all these questions is expected to be negative.
This proposal creates a new, additional service that completes the alternatives
that musicians already use. An analogue synthesiser online does not replace the
experience of having a real keyboard to play with, listen to, learn from, know how
to programme or play live. Many musicians can afford to have dedicated, analogue
and vintage machines, and the service does not replace them. But they will be
possibly interested in accessing many others made available from a synthesiser farm.
This should also increase the interest in software plug-ins to test and arrange the
production before asking the tracks to be rendered by real machines. Even people
that own synthesisers may find useful to ask the automated system to record the
final track with the very same synthesiser but with a superior recording process.
Therefore, the availability of such a service can provide additional reasons to seek
for analogue synthesisers and plug-ins, helping to increase the overall market.

5.3.4   Simple Cost Estimation

As it was said before, a key issue for the system's economic viability is the high
scalability attainable with the batched access. It is yet to be verified whether users
will adapt themselves to the system operation or not, but it certainly depends on
the accessing costs. If we can prove that the cost for accessing the service is low,
it is expected that people will recognise the possibilities, the aggregate value and
the returns in terms of quality that they can get from the system and use it. Let us
assume that the market exists and is larger than what can be provided by a single
synthesiser of a given kind. Based on this assumption, we find that the system should
be sized according to the demand so as to keep its utilisation bounded by some
parameter. Assuming a basic M/M/1 queue model, for simplicity, with utilisation
rate $\rho = \lambda/\mu$ of 50 %, from basic queue theory, we get that the number of clients
in the queue is expected to be on average $N = \rho/(1 - \rho)$ or 1 client. Therefore,
for a given average job size, the client is expected to wait twice this time (including
the time to process the 1 client in the system and its own job), which is a good
quality of service. Assume now that a US$2,500.00 synthesiser is being operated
by a computer system and accompanying audio/MIDI interfaces, also in the total
value of US$2,500.00. Now let us suppose that the system will charge only US$1.00
for each minute of sound played and recorded. At 50 % of utilisation, the system
should operate 720 min/day, generating a revenue of around US$500 (remember that

there is the overhead of the delays when processing each job). In such hypothetical scenario, the equipment would pay by itself in just 10 days. While we are ignoring the other costs of operating the service, the above numbers were presented to show that a US$100/min target price is not unrealistic. Now by comparing this price (and the easy web access) to the option of renting a keyboard or renting a high-quality recording studio, that can reach hundreds of dollars per hour, to do the same job, it becomes clear that this is an attractive alternative to many users. We can expect that a lot of people that would never be able to use the standard alternatives will try the new service.

5.3.5   Ubiquitous Access with Uncompromised Quality

The modification of instruments is another possible advantage of the proposed model. The technicians of top artists occasionally work with instrument makers or implement custom modifications by themselves that improve on the quality of a particular instrument compared to the outcomes of a large-scale production. In other words, it is possible to employ both custom recording hardware and custom instruments. The possibility comes from the fact that in large-scale production, there is always a compromise between price and performance, yet to implement a single server unit, non-compromised components may be employed, as the scale comes from the higher utilisation. The system can allow widespread and low-cost access to systems built with no compromise, with the best sonic performance, not attainable in units from large-scale production.

A system like this extends the notion of Ubiquitous Music to include access to analogue instruments. The research areas of Ubiquitous Music and computer music usually employ modern interfaces in the form of new hardware designs or consumer products as new ways to make music. But the sound generation is usually either locally produced or digitally simulated. In our proposal, we open the possibility to accessing analogue and real instruments remotely from anywhere, no matter which kind of device is being used. In a simple example, a person can create a music while operating his cell-phone or tablet in a bus, sending the notes to be played with a vintage analogue synthesiser and listening to the resulting sound in just a few seconds or minutes. The existence of analogue servers can help promote the use of analogue synthesisers, increase public knowledge about them, increase the need for modelling plug-ins and provide funds for the conservation of older instruments in operating conditions. Finally, it is speculated that a similar approach can be employed to access real acoustic instruments, as it will be considered in the next sections.

5.3.6 Other Business Models

In the preceding discussion, we have been considering the commercial implementation of a system that provides the service and considering that it should
be technically and economically viable for both end users and the company that
operates it. A good set of arguments was already provided to support the advantages
on both sides, with plenty of aggregate value to explore. There is however other
possible business models that shall be at least briefly mentioned here.

First of all, like in many other web or cloud services available to broad audiences,
the service can indeed be provided for free and paid for in terms of advertisement
or other aggregate values. In the case of synthesisers, specifically, it can be tied
and explored by both physical instrument manufacturers and by software plug-
in vendors. In the later case, the service of accessing the analogue server can be
associated with the acquisition of a state-of-the-art software plug-in. In this case,
the user has an additional strong appeal to buy such software. The user knows
that he/she has the software version in real time at its disposal anytime and also
the possibility of leveraging the quality whenever he/she is going to record a piece
for an album release by remotely accessing the real instrument, closely tuned and
matched. In the case of manufacturers, they may want to finance the operation of
a set of their products as servers in order to promote their brand, specific models,
making everyone aware of the instrument so that people get used to them and want
to buy the real instruments, increasing their market.

Yet another way of accessing analogue synthesisers online would be to create
a network of end users, implementing the service in a peer-to-peer approach. In
this model, the system is free from costs, being composed by the aggregated set of
machines online provided by the users themselves. In this case, there is no guarantee
about the availability of the machines, as the users can turn their systems on and off
whenever desired. In fact, they will be eventually playing with their instruments,
and they have absolute priority to do that, locking their systems frequently. On the
other hand, as expected, an owner only uses an instrument in a very small fraction of
the time, with few exceptions. A simple mechanism of credits may help ensure that
users will turn their systems on and log in at the network to contribute and further
benefit from it.

The uncertainty in terms of available machines can potentially be compensated
by the number of users. If a synthesiser is popular, many owners may be registered
in the system, what can potentially ensure that there will always be one or some of
them operating online. As any other peer-to-peer system, it strongly relies on the
number of registered users, or its popularity, for properly working.

The major challenge to create a peer-to-peer system however would be the
installation procedures and requirements in terms of hardware and software,
specially considering the desired audio quality, which is one of the most important
aspects, closely related to the motivation of high quality sound rendering. If user
has a low-cost audio interface, a noisy computer or MIDI interface exhibiting
significant jitter, the recorded audio might not achieve the desired quality, becoming

useless. On the other hand, provided that the service is free of charge, the user can submit the job again to another peer, and the system may find its equilibrium. As regarding the installation procedures, the variety of hardware and software systems and configurations may become difficult to address to reach a significant number of users. Nevertheless, this is indeed an interesting implementation option, one that can grow with the community, using open-source software and decentralised concepts.

5.4 Audio Processes Suitable for Access as Remote Servers

The main purpose of the remaining of the chapter is to verify which other audio processes can be treated as analogue or acoustical servers for remote and batched access targeting high-quality sound rendering. We also want to check the needed technology and the challenges that must be undertaken. The first step, which justifies and defines the contribution of the present work, is to understand that there are many audio processes or activities that cannot be performed remotely, in principle.

5.4.1 What Cannot Be Done

Starting with the processes or equipment, we can easily see that the following individual items cannot be used remotely: microphones and preamplifiers. It is not possible to have a remote microphone available for use, because it is not possible to mechanically transmit the sound to be captured without being previously captured by a microphone.

That same simple principle applies to the preamplifier need to increase the tiny microphone signal to appropriate power/voltage levels that run regular electronic audio, called line level. In practice, good studios have a large set of different microphones and preamplifiers with different technologies and sound signatures. Therefore, those two items alone represent a significant part of the studio equipment.

Guitar cabinets or preamplifiers lie in a somewhat grey category. As they are actual preamplifiers, needed to increase the power of small signals, at first it may be difficult to capture the small guitar signal with accurate analogue-digital conversion to transmit it into a preamplifier remotely located. But considering that guitars may have already an internal preamplifier or the sound may pass through effect pedals, it may indeed be possible to design circuits that capture it, record and then use the cabinet or last preamplifier of the path in a remote server. This can be hard to set up; it may work just under some circumstances or may be considered just as a post-recording sound effect. It is, however, worth to mention that the musician could play using a local, low-quality guitar cabinet but with a system that is also recording the unprocessed (low power) signal to be feed through a higher-quality cabinet later on.

Last, there are activities that would not be done remotely in a batched system because it does not make sense. The previous example of guitar amplifiers might fall into this category, as in many cases this type of music is intrinsically dependent on the sound that the musician is experiencing. For an example, a sensible performance that explores microphonic guitar tones is only possible with the actual equipment in front of the musician. Any activity that depends on fine nuances of the actual equipment will have the same situation. Another aspect is that many musical activities must be carried out instantaneously; the fine nuances of the actual process must be immediately available. The musicians usually need to feel the actual instrument in its full richness to be able to perform well, as music is a sensible art form, whose interpretation depends both on technique and emotion.

Having identified the few things that are not possible or are inappropriate for remote use, it becomes clear that many things are indeed possible. In all the cases we are considering the careful application of the concept of separate preview and final sound rendering. In other words, it would be only worthwhile to implement remotely processes that benefit a lot from the specialisation a server can afford and for which an "economic" preview can be provided with the required accuracy. Let us then analyse such possibilities.

5.4.2 More About Synthesisers

The first prototyped system described in Sect. 5.2 shows that it is quite straight-forward to put a modern MIDI-equipped analogue synthesiser online. Those are the first options to be considered, of course. There are many models of analogue synthesisers in production today, with MIDI. Monophonic synthesisers and analogue modules comprise the majority of the systems in production, while polyphonic machines, like the Alesis Andromeda, were almost all discontinued. Regarding their prices, while there is a set of analogue synthesisers that sell for under a thousand dollars, they cannot be considered cheap. Those are the simpler generators, with limited resources, monophonic, and they are usually good in a specific type of sound, as most of the instruments. Therefore, no musician would be complete buying just one of those instruments. It has to be at least the second instrument and will find a rather limited usage, turning a modest-looking investment into an expensive one.

While it seems simpler to implement a service using instruments currently in production, the system may help to preserve the entire sound palette that many generations of synthesisers could create by promoting the continuous use of real analogue synthesisers of all times. The synthesiser must be MIDI playable, but many synthesisers that are not MIDI capable can be controlled with specific hardware. There are standard MIDI-to-CV converters (CV stands for control voltage) that can be used to play older analogue synthesisers that do not support MIDI directly. Even synthesisers that do not have CV inputs and cannot be programmed remotely can

Table 5.1 A small sample of current and old analogue synthesisers for sale

Model	Manufacturer/brand	Price (U$)	Polyphony	In production?
Voyager Rck	Moog Music	$2,795.00	Monophonic	Yes
SE-1X	Studio Electronics	$1,679.00	Monophonic	Yes
Omega8	Studio Electronics	$4,999.00	Polyphonic	Yes
Prophet five	Seq. circuits	$4,258.00	Polyphonic	Used
CS80	Yamaha	$18,249.00	Polyphonic	Used
Jupiter8	Roland	$2,699.00	Polyphonic	Used
Odyssey	ARP	$2,529.00	Monophonic	Used

The price for used models was obtained from Ebay on May 20, 2012

be automated with a dedicated hardware work on them. With hardware work, it is also possible to make some instruments programmable, where they originally were not. There is a difference in just playing them through MIDI-to-CV interfaces or allowing different sounds to be programmed and accessed. The last level of integration would be to allow re-patching of modular systems, where the sounds must be configured by patch cables. This is naturally also possible with a set of automated relays.

The importance of automating, preserving and letting more people access those instruments transcends the commercial application of the system and must be considered from a cultural and historical perspective. This investment might be key to help keep iconic instruments alive and working for future generations in an affordable and cost-effective way. It can be seen that old machines in good condition have been increasing in prices. In Table 5.1, a small sample of some old machines that are available for sale on Ebay has been listed, together with some models in production. It can be observed that there are available units for the most popular synthesisers, as a consequence that they have been produced in a good scale in the past. On the other hand, theirs prices are relatively high. Those prices were averaged on the available listings on May 20, 2012, and can be compared to what is available in other records [18, 22] from the past, exhibiting an expressive increase.

5.4.3 Acoustic Instruments

Acoustic instruments are very different from electronic ones and require special setups to be automated, but there have been many successful works in the area—mixed with the field of robotics and artificial intelligence—like, for example, a robot percussionist that reacts to a human player [28]. In "A History of Robotic Instruments" [13] we find an excellent prospection of the field and numerous acoustic instruments that are currently passive of automation. Some of these contraptions have even found their way into more mainstream music venues when Eric Singer and the LEMUR group [24] supplied Jazz musician Pat Metheny with his very own robotic band for the album Orchestrion [16]. Most of these robotic instruments are MIDI controlled and actuate with the help of motors and solenoids. This scenario shows itself promising to the ambitions of analogue recording servers.

Two obvious choices of instruments to be considered at first are the piano and the Hammond organ. The widespread use of both is an indication that there will be high demand for their use. Since the appearance of the piano, the instrument has captured the attraction of most of the music genres. It is equally used in traditional and modern classic concerts, for jazz and blues, popular music, rock, as well as many local music styles. Its importance at the same time mean that it will be well used if made available online and also signifies that different types of piano sound exist that would be appropriate for specific purposes. Despite of that, its automation does not have the parameter preview problem that synthesisers have. The user will still need sound preview to select one piano if many are available, but there are no parameters to set. Pianos are maybe one of the oldest forms of automated instruments. Piano rolls were rolls of perforated papers, a format introduced before 1900 that could execute pre-recorded performances in special pianos.

Today there are alternatives like the Disklavier [3, 4]. The reliability of the Disklavier and the Bösendorfer SE for expressive playing reproduction (recorded as a MIDI file) is analysed in [8,9]. The pianos were found unable to precisely emulate recorded tempo and dynamics, for instance, *soft notes* are played more loudly than they were originally recorded. Nevertheless the papers [8, 9] states that the tempo changes are in the order of less than 50 ms and that in the course of the testing, the majority of MIDI notes velocity were between 40 and 80, and both pianos had no problem reproducing them accurately. Unfortunately there are no studies addressing the reproduction of the newer Bösendorfer CEUS model, only tangential works as, for example, [26] that suggests it is capable of recording expressive playing in a high-resolution notation.

With those instruments, there is already available technology to operate a system based on the remote recording of pianos and achieve a reasonable result in terms of performance accuracy. In fact it has come to our attention that a remote recording system, very similar to the proposed herein, is being offered in the SparkWorld Studio [25], which provides high-quality recordings with a Yamaha Disklavier Piano. They observe that by operating their studio as a "piano factory" for a few days in a month, it is possible to provide lower recording costs. On the other hand, it must be noted that their entire process is ad hoc, under human supervision and control. The piano is automated, but the job processing and recording are not. Their system is therefore dedicated to only professional musicians and does not characterise a more accessible Ubiquitous Music approach.

The operation of the SparkWorld piano in part time is the first strong indication of the available market and demand what we could find. On the other hand, there is a new requirement to implement such a service in a larger scale, the problem of tuning. Piano tuning is not only a delicate and time-consuming task but also takes the form of art. Different tuning systems exist, and a piano technician is often required to prepare the instrument to a particular composition and performer. Nevertheless, in a recent paper [10], a system was developed that better explains the many particularities and deviations that a human technician does, and it is able to mathematically optimise the piano tuning based on our psychoacoustic perception of the interaction among all harmonic frequencies in a particular instrument and particular room. While the tuning is usually performed manually, in such an automated instrument, it is possible to develop special hardware that mechanically adjusts the instrument from time to time, and there is at least a patent already developed for such a system [7].

The "miking", e.g. microphone selection and placement, is of paramount importance, as it will be for other acoustical instruments. A grand piano has a lid that must be open to better amplify and reflect the sound to the audience. As an example of sound capture option, an automated piano can be recorded with the lid closed, open or detached, so as to avoid direct and reflect sounds to mix in different phases. This is just one of the many options and considerations to address. An automated system must of course be built on top of the best practices already used for recording pianos in today's studios. A set of different microphones, as well as some placement adjustments, can be provided to get a variety of sound perspectives, and of course the most popular are the ones to be covered. Despite the challenges, as an educated guess we should say that a piano can be one of the most valuable instruments to offer online. There are plenty of modest home users that would be able to compose in their computers and then test their ideas with the sound of the real thing.

Hammond organs, being electrically controlled, can be modified to respond to MIDI messages, including parameters of the performance, such as the Leslie speaker rotation. Different levels of automation can be devised, for example, the user may be able to choose from the organ presets or in a more complex setup adjust any one of the instrument drawbars. The sound capture must be done taking in consideration the Leslie speaker. The electromechanical system on those organs is also beneficial as there is little overhead between performances, e.g. there is no need to re-tune the instrument.

Traditionally, pipe organs are enormous instruments located in churches or theatres and that has natural implications in their cost and difficulty of access. There are, however, smaller versions of pipe organs, like the chamber organs, and those could be attractive options to be considered for first automation projects. Recently there have been efforts to install MIDI receivers in these instruments [5] and some of those are even at people's disposal, for example, the massive pipe organ at Melbourne Town Hall [2]. The challenge lies in the allocation of a space that can accommodate the instrument and the automation system. The cost of such a setup could however be viable if such a system were to be employed and permitted a maximum utilisation of the instrument.

5.4.4 Analogue Mixing

Until the mid-1980s, recording studios used pure analogue recording processes, storing the sound signals in magnetic tapes. The very expensive tape machines represent the culmination of that technology, improved further by dynamic compression optimised to allow higher dynamics and separate the signal from the medium noise. While major studios still keep and occasionally use those tape machines today, digital technology replaced the majority of the storage and manipulation tasks. In the current scenario, for tracking or sound capture, the signal comes from microphones or directly from electronic instruments, passes through preamplification, is converted into the digital domain by an analogue–digital converter (ADC) and then stored and manipulated easily in the DAWs.

The mixing activity is defined as the correct composition of the individual tracks with their relative volumes, panoramic positions, effect sends and other serial processing and is both a technical and a decisive artistic task [11]. For it to take effect, the individual tracks must be added together. Both the mixing and the addition itself can be done inside the computer (in the DAW) our "out of the box". In smaller studios it is usually done inside the DAW for cost reasons, while large studios use a combination of operations in the digital domain together with the access of a big external mixing console. With this option of analogue mixing in the external console, there are additional DA and AD conversion steps. While theoretically that could make the audio worse, the opposite is frequently considered as true, although there is much controversy as to why.

A simple understanding can be achieved if we interpret the mixing as another part of the art composition [23], where the mixer along with the other analogue gear has a role as important as a musical instrument. In this sense, it is easier to accept that an analogue mixing can sound subjectively better than a correct addition in a computer. Anyway, all those analogue equipments took decades to evolve and be selected as good tools to make music. As a result, many people seek for the sound of analogue mixing, and there are small, dedicated units to be used just for the analogue "summing" in productions that do not have access to large mixing consoles. Despite of that, they are also expensive and must be combined with expensive DA and AD converters, for obvious reasons.

Therefore, considering that the final mixing is another analogue process already present in many production's signal path and that there is much search for its subjective characteristics, it is as a good candidate to be offered as an automated server. There are a few differences and challenges to implement it, as described below. The first technical difference comes from the amount of information to be considered, specially transferred between the user and the server. While MIDI tracks contain just a few kbytes and the final recording of an analogue synthesiser will contain part of the duration of the song possibly in mono, the mixing activity requires that all tracks be individually transferred to the server and the resulting stereo or multi-channel mix be transferred back in full. Consider a basic 24-track production in 24 bit/48 kHz, in a 4 min song to mix down. That would require 829 MB to be transferred to the server and approximately 69 MB to be retrieved. While it does not seem too much traffic for today's parameters, and from the point of view of the user, we must emphasise that this was the amount of information estimated for just 4 min of processing in the mixing server, and a continuous flow will be needed for high-capacity operation. So appropriate care must be taken to implement the system communication mechanism. There are many alternatives to transfer the audio tracks. The most simple would be to get independent audio files for each track in a standard format, all with the duration of the song. That would use much more information, as tracks are not usually sounding from the beginning to the end. But this simple interface makes the system independent of the DAW software and versions. On the other hand, it is possible to transfer the project files, and that may save size if the project is consolidated and unused takes are deleted. It also requires the service structure to acquire, support and start up different DAWs, what can be both expensive and complex to maintain.

The next issue is related to automation. It is necessary to implement what is called "mixing automation", which is the dynamic determination of mixing parameters during the composition. The automation is used to implement volume changes and, panoramic effects and is nowadays recorded as if it was another performance during the song. There are many options as to how the automation could be employed in a remote server, and we must consider some of their implications carefully.

As a first option, the automation can be done digitally, inside the DAW, so that the actual mixing server has fixed settings. This is, however, the worst option. While the computations inside most of the DAWs use floating-point arithmetic, the DAC interfaces (as all audio interfaces) employ fixed-point numbers, and reducing the volume of the tracks in the digital domain, as usually needed, will reduce the resolution accordingly. It must be checked or investigated further in a real production facility if this fact will be perceived as loss of quality, but it is indeed worse than the DAW mixing and worse than an analogue volume, technically speaking.

The other alternative is to read the DAW automation data and reproduce it externally in the server's mixing console. There is already technology used to do that implemented by major mixing console builders. In fact, this integration between digital and analogue is already widely used. Our proposal is to make such integration independent of the location. There is also the potential need to access such data without being tied to specific companies that implemented it as proprietary technology. One must also be able to reproduce it to implement dedicated mixers that differ from the ones provided by existing companies, as addressed below. Besides the technology itself, there is also the question of DAW integration. Whatever the solution is best or just selected, it must be easy to integrate it with existing DAWs, except if the DAW's company is directly involved in the implementation, in which case everything can be adapted.

One example of such question is how to get the original, unprocessed audio, even when the project is being listened by the user (prototyped) with the final automated parameters in its DAW. The problem comes from the fact that we need to operate seamlessly in a different setup. If the user has an automated mixing desk, the audio goes out unprocessed along with the controlling commands (it is processed at the desk), but this is the user that does not need to access the remote server, in principle. Yet the user that does not have an automated mixing desk makes all automation in the DAW, so the audio is automatically already processed by the DAW. Even if the DAW routing allows it internally to send the unprocessed audio out, it might be complicated to set it in each user project, and it may discourage people from using it. An alternative way of integrating that would be through a dedicated mixer plug-in. As software plug-ins can receive an arbitrary number of audio and MIDI channels, the remote mixing server can be integrated as a dedicated mixing plug-in, relatively simplifying the user perception.

There is a very important matter regarding mixer quality, type, size and general options, closely associated with application and aesthetics. From a purist point of view, the analogue mixer should sound as transparent as possible, while maintaining musicality. Classical music would normally call for such cleanest paths, employing class A tube designs with few stages. On the opposite hand, popular music used to prey the sound of vintage transistorised consoles, with their characteristic sound that came from audio transformers, old resistors, point-to-point connections and so on. Therefore, one must consider different mixing machines targeted for different applications and sound flavours. The market itself will strongly call to use machines that are already known, so that, as an example, a vintage Neve mixer online would have much more market appeal than a brand new and very clean design.

5.4.5 Effect Processors

Finally, there are many effect units that are employed mostly, but not necessarily, during mixdown, and some of them may be expensive vintage analogue units that could be made available as audio servers. Typical examples are dynamic processors (compressors, expanders, limiters and noise gates), specially those which employ high-quality transformers, tubes optical parts. Some reverb units, both digital and electromechanical, like plate reverbs, together with many other audio effects that transform the sound and are used to process or complement the musical tracks, also fall in this category. If the units are used only once at mixdown, they can be integrated in the mixing service, and would not benefit from being available as separate servers. Only after basic functions are used and tested in the system it will become clear which other effects are demanded and can benefit from the new model.

Conclusions

This chapter described the concept of remote analogue audio recording, which is an innovative usage model for analogue and acoustical equipment and provided a prospective analysis of possible instruments and operations that can be implemented as analogue audio servers without real-time access. In the proposed access model, real instruments are installed as servers and accessed sequentially for processing jobs submitted by the users to obtain high-quality

(continued)

audio recordings. Such a system allows ubiquitous access to real instruments, from any device that can connect to the Internet.

The proposed model has many advantages. It can provide low accessing cost for end users even when employing expensive and specialised hardware, because of the high utilisation rates attainable with a batched system that minimises idle time. It can increase the variety of instruments that both novices and professionals can use, by operating collections of instruments from different types, brands, classes, and so on. The system can help to restore and maintain historical instruments, as well as custom and unique instruments created for community access. The proposed model works together with other options. Users that have good instruments can access others. Those that cannot afford to buy any instrument can access them remotely. Musicians that are used to work with virtual instruments may continue to do so and benefit for an improved quality as they ask for a final rendering with real instruments. The culture and knowledge about real instruments and analogue synthesisers can be improved as people are allowed to access them from wherever they are.

Many other audio processes besides the analogue synthesisers initially considered can benefit from the operation as audio servers, and we provided an initial assessment of which are the required technologies and challenges. There seems to be a very broad range of instruments and audio processing possibilities for such a system. Many kinds of synthesisers, acoustic instruments and even mixing consoles can be remotely offered, and we found that there is available technology to operate most of those servers. Together, that should help promoting Ubiquitous Music creation while preserving the historical palette of music instruments and processes.

The concept of separating the sound preview from the final sound rendering is a key component that helps to reduce the cost and keep flexibility. In this way, users can focus on local music composition, recording and production, listening to sound preview, without requiring continuous live access to a remote server, and later on access high-quality recording for final sound rendering. The availability of musical software and processing power in consumer devices like cell-phones, tablets and laptops is more than ever shifting the focus of musical activities from consumption to composition and production. Our contribution lies in bringing the world of real instruments, analogue processes and high-quality recordings closer to these devices by means of ease of access and low cost, effectively making these processes ubiquitous.

References

1. Cáceres, J., Chafe, C.: JackTrip: under the hood of an engine for network audio. J. New Music Res. **39**(3), 183–187 (2010). doi:10.1080/09298215.2010.481361
2. City of Melbourne: Town Hall Organ. http://www.melbourne.vic.gov.au/ (2012)
3. Corporation, Y.: Bösendorfer CEUS Recording System. http://www.boesendorfer.com/index.php?m=70 (2012). Accessed 5 April 2012
4. Corporation, Y.: Yamaha Disklavier Piano. http://usa.yamaha.com/products/musical-instruments/keyboards/disklaviers/ (2012). Accessed 7 April 2012
5. Edwards, G.: Computer Interface Makes 19th-Century Pipe Organ Rock. Xcell J. **67**(First quarter 2009), 44–49 (2009)
6. Electronics, S.: Studio Electronics ATC-1. http://www.studioelectronics.com/ (2004). Accessed 5 April 2012
7. Gilmore, D.A.: Automatic piano tuner (1998). US Patent 5,756,913
8. Goebl, W., Bresin, R.: Are computer-controlled pianos a reliable tool in music performance research? Recording and reproduction precision of a Yamaha Disklavier grand piano. In: Proceedings of the Meeting on Music Performance: Analysis, Modeling, Tools. University of Padova, Padova, Italy (2001)
9. Goebl, W., Bresin, R.: Measurement and reproduction accuracy of computer-controlled grand pianos. J Acoust. Soc. Am. **114**(4), 2273 (2003). doi:10.1121/1.1605387
10. Hinrichsen, H.: Entropy-based tuning of musical instruments. Revista Brasileira de Ensino de Física **34**(2), 1–8 (2012)
11. Izhaki, R.: Mixing Audio: Concepts, Practices and Tools. Taylor & Francis, London (2008)
12. Jensenius, A.R., Castagn, N., Camurri, A., Maestre, E., Malloch, J., Mcgilvray, D.: A Summary of Formats for Streaming and Storing Music-Related Movement and Gesture Data. In: Proceedings of the 4th International Conference on Enactive Interfaces (ENACTIVE/07). Grenoble, France (2007)
13. Kapur, A.: A history of robotic musical instruments. In: Proceedings of the International Computer Music Conference (ICMC). Barcelona, Spain (2005)
14. Kapur, A., Wang, G., Davidson, P., Cook, P.R.: Interactive network performance: a dream worth dreaming? Organised Sound **10**(03), 209 (2005). doi:10.1017/S1355771805000956
15. Mayton, B., Joliat, N., Paradiso, J.A.: Patchwerk: multi-user network control of a massive modular synthesizer. In: Proceedings of the Conference on New Instruments for Musical Expression (NIME'12), University of Michigan, Ann Arbor (2012) In: NIME Proceedings (2012)
16. Metheny, P.: Orchestrion. Warner Music (2010)
17. Moog Music Inc.: http://www.moogmusic.com/ (2012)
18. Oppenheimer, A.: Used Gear Price List. http://archive.cs.uu.nl/pub/MIDI/DOC/prices-used-USA. Accessed 30 May 2012
19. Orto, C., Karapetkov, S.: Music Performance and Instruction over High-Speed Networks. Tech. Rep. June, polycom (2011). http://www.polycom.com/global/documents/whitepapers/music_performance_and_instruction_over_highspeed_networks.pdf
20. Pimenta, M.S., Flores, L.V., Capasso, A., Tinajero, P., Keller, D.: Ubiquitous music: concept and metaphors. In: Farias, R.R.A., Queiroz, M., Keller, D. (eds.), Proceedings of the Brazilian Symposium on Computer Music (XII SBCM), pp. 139–150. SBC, Recife, PE (2009)
21. Pinch, T., Reinecke, D.: Technostalgia: How Old Gear Lives on in New Music. In: Bijsterveld, K., Van Dijck, J. (eds.), Sound Souvenirs: Audio Technologies, Memory and Cultural Practices, pp. 152–166. Amsterdam, University Press, Amsderdam (2009)
22. Prepal: Internet Used Musical Instruments. http://www.prepal.com/ (2004). Accessed 30 May 2012
23. Radanovitsck, E.A.A., Keller, D., Flores, L.V., Pimenta, M.S., Queiroz, M.: mixDroid: Time tagging for creative activities (mixDroid: Marcação temporal para atividades criativas). In: Costalonga, L., Pimenta, M.S., Queiroz, M., Manzolli, J., Gimenes, M., Keller, D., Farias,

R.R. (eds.), Proceedings of the XIII Brazilian Symposium on Computer Music (SBCM 2011). SBC, Vitória, ES (2011).
24. Singer, E., Larke, K., Bianciardi, D.: LEMUR GuitarBot: MIDI robotic string instrument. In: Proceedings of the 2003 Conference on New Interfaces for Musical Expression, pp. 188–191. National University of Singapore, Singapore (2003)
25. Studio, S.: Sparkworld Studio. http://www.sparkworld.com (2012). Accessed 30 April 2012
26. Traube, C.: Piano touch analysis: a matlab toolbox for extracting performance descriptors from high-resolution. In: Actes des Journées d'Informatique Musicale (JIM 2012), pp. 9–11. Mons, Belgium (2012)
27. Valladares, T.A.: Independent Producers: A Guide to 21st Century Independent Music Promotion and Distribution. Master's Degree Thesis. Univeristy of Oregon (2011)
28. Weinberg, G., Driscoll, S.: The interactive robotic percussionist: new developments in form, mechanics, perception and interaction design. In: Proceedings of the ACM/IEEE international conference on Human-robot interaction, pp. 97–104. ACM, New York (2007)
29. Zawacki, L.F., Johann, M.O.: A system for recording analog synthesizers with the Web. In: Proceedings of the International Computer Music Conference, pp. 128–131 (2012)

Part III
Technology

Chapter 6
Development Tools for Ubiquitous Music on the World Wide Web

Victor Lazzarini, Edward Costello, Steven Yi, and John ffitch

Abstract This chapter discusses two approaches to provide a general-purpose audio programming support for Ubiquitous Music web applications. It reviews the current state of web audio development and discusses some previous attempts at this. We then introduce a JavaScript version of Csound that has been created using the Emscripten compiler and discuss its features and limitations. In complement to this, we look at a Native Client implementation of Csound, which is a fully functional version of Csound running in Chrome and Chromium browsers.

6.1 Introduction

The web browser has become an increasingly viable platform for the creation and distribution of various types of media computing applications [11]. It is no surprise that audio is an important part of these developments. For a good while now, we have been interested in the possibilities of deployment of client-side Csound-based applications, in addition to the already existing server-side capabilities of the system. Such scenarios would be ideal for various uses of Csound. For instance, in Education, we could see the easy deployment of computer music training software for all levels, from secondary schools to third-level institutions. For the researcher, web applications can provide an easy means of creating prototypes and demonstrations. Composers and media artists can also benefit from the wide reach of the internet to create portable works of art. In summary, given the right conditions, Csound can provide a solid and robust general-purpose audio development environment for a variety of uses. In this chapter, we report on the progress towards supporting these conditions.

V. Lazzarini (✉) • E. Costello • S. Yi • J. ffitch
Music Department, Maynooth University, Ireland
e-mail: vlazzarini@nuim.ie; edwardcostello@gmail.com; stevenyi@gmail.com;
jpff@codemist.co.uk

© Springer International Publishing Switzerland 2014
D. Keller et al. (eds.), *Ubiquitous Music*, Computational Music Science,
DOI 10.1007/978-3-319-11152-0_6

111

6.2   Audio Technologies for the Web

The current state of audio systems for World Wide Web applications is primarily based upon three technologies: Java[1] Adobe Flash,[2] and HTML 5 Web Audio.[3] Of the three, Java is the oldest. Applications using Java are deployed via the web either as applets or via Java Web Start. Java as a platform for web applications has lost popularity since its introduction, primarily due to historically sluggish start-up times as well as concerns over security breaches. Also of concern is that major browser vendors have either completely disabled applet loading or disabled them by default and that NPAPI plugin support, with which the Java plugin for browsers is implemented, is planned to be dropped in future browser versions. While Java sees strong support on the server side and desktop, its future as a web-deployed application is tenuous at best and difficult to recommend for future audio system development.

Adobe Flash as a platform has seen large-scale support across platforms and across browsers. Numerous large-scale applications have been developed such as AudioTool,[4] Patchwork[5] and Noteflight.[6] Flash developers can choose to deploy to the web using the Flash plugin, as well as use Adobe Air[7] to deploy to desktop and mobile devices. While these applications demonstrate what can be developed for the web using Flash, the Flash platform itself has a number of drawbacks. The primary tools for Flash development are closed-source, commercial applications that are unavailable on Linux, though open-source Flash compilers and IDEs do exist.[8] There has been a backlash against Flash in browsers, most famously by Steve Jobs and Apple,[9] and the technology stack as a whole has seen limited development with the growing popularity of HTML 5. At this time, Flash may be a viable platform for building audio applications, but the uncertain future makes it difficult to recommend.

Finally, HTML 5 Web Audio is the most recent of technologies for web audio applications. Examples include "Recreating the sounds of the BBC Radiophonic

---

[1]http://java.oracle.com.

[2]http://www.adobe.com/products/flashruntimes.html.

[3]http://www.w3.org/TR/webaudio/.

[4]http://www.audiotool.com/.

[5]http://www.patchwork-synth.com.

[6]http://www.noteflight.com.

[7]http://www.adobe.com/products/air.html.

[8]http://www.flashdevelop.org/.

[9]http://www.apple.com/hotnews/thoughts-on-flash/.

Workshop using the Web Audio API" site,[10] Gibberish[11] [9] and WebPd.[12] Unlike Java or Flash, which are implemented as browser plugins, the Web Audio API is a W3C proposed standard that is implemented by the browser itself.[13] Having built-in support for Audio removes the security issues and concerns over the future of plug-ins that affect Java and Flash. However, the Web Audio API has limitations that will be explored further below in the section on Emscripten.

6.3   Csound-Based Web Application Design

Csound is a music synthesis system that has roots in the very earliest history of computer music. Csound use in desktop and mobile applications has been discussed previously in [7, 8, 13].

Prior to the technologies presented in this chapter, Csound-based web applications have employed Csound mostly on the server side. For example, NetCsound[14] [5] allows sending a CSD file to the server, where it would render the project to disk and email the user a link to the rendered file when complete. Another use of Csound on the server is Oeyvind Brandtsegg's VLBI Music,[15] where Csound is running on the server and publishes its audio output to an audio stream that end users can listen to. A similar architecture is found in [6]. Since version 6.02, Csound also includes a built-in server that can be activated through an option on start-up. The server is able to receive code directly through UDP connections and compile them on the fly.

Using Csound server side has both positives and negatives that should be evaluated for a project's requirements. It can be appropriate to use if the project's design calls for a single audio stream/Csound instance that is shared by all listeners. In this case, users might interact with the audio system over the web, at the expense of network latency. Using multiple real-time Csound instances, as would be the case if there was one per user, would certainly be taxing for a single server and would require careful resource limiting. For multiple non-real-time Csound instances, as in the case of NetCsound, multiple jobs may be scheduled and batch processed with less problems than with real-time systems, though resource management is still a concern.

An early project to employ client-side audio computation by Csound was described in [3], where a sound and music description system was proposed for the rendering of network-supplied data streams. A possibly more flexible way to use

---

[10]http://webaudio.prototyping.bbc.co.uk/.

[11]http://github.com/charlieroberts/Gibberish.

[12]http://github.com/sebpiq/WebPd.

[13]See http://caniuse.com/audio-api for browser support.

[14]http://dream.cs.bath.ac.uk/netcsound/.

[15]http://www.researchcatalogue.net/view/55360/55361.

Csound in client-side applications, however, is to use the web browser as a platform. Two attempts at this have been made in the past. The first was the now-defunct ActiveX Csound (also known as AXCsound),[16] which allowed embedding Csound into a web page as an ActiveX Object. This technology is no longer maintained and was only available for use on Windows with Internet Explorer. A second attempt was made in the Mobile Csound Project [8], where a proof-of-concept Csound-based application was developed with Java and deployed using Java Web Start, achieving client side Csound use via the browser. However, the technology required special permissions to run on the client side and required Java to be installed. Due to those issues and the unsure future of Java over the web, the solution was not further explored.

The two systems described in this chapter are browser-based solutions that run on the client side. The both share the following benefits:

- Csound has a large array of signal processing opcodes made immediately available to web-based projects.
- They are compiled using the same source code as is used for the desktop and mobile version of Csound. They only require recompiling to keep them in sync with the latest Csound features and bug fixes.
- Csound code that can be run with these browser solutions can be used on other platforms. Audio systems developed using Csound code is then cross-platform across the web, desktop, mobile and embedded systems (i.e. Raspberry Pi, Beaglebone, discussed in [1]). Developers can reuse their audio code from their web-based projects elsewhere and vice versa.

6.4 Emscripten

Emscripten is a project created by Alon Zakai at the Mozilla Foundation that compiles the assembly language used by the LLVM compiler into JavaScript [14]. When used in combination with LLVM's Clang frontend, Emscripten allows applications written in C/C++ or languages that use C/C++ runtimes to be run directly in web browsers. This eliminates the need for browser plugins and takes full advantage of web standards that are already in common use.

In order to generate JavaScript from C/C++ sourcecode, the codebase is first compiled into LLVM assembly language using LLVM's Clang frontend. Emscripten translates the resulting LLVM assembly language into Javascript, specifically an optimised subset of JavaScript entitled asm.js. The asm.js subset of JavaScript is intended as a low-level target language for compilers and allows a number of

---

[16]We were unable to find a copy of this online, but one is available from the CD-ROM included with [2].

optimisations which are not possible with standard JavaScript.[17] Code semantics which differ between JavaScript and LLVM assembly can be emulated when accurate code is required. Emscripten has built-in methods to check for arithmetic overflow, signing issues and rounding errors. If emulation is not required, code can be translated without semantic emulation in order to achieve the best execution performance [14].

Implementations of the C and C++ runtime libraries have been created for applications compiled with Emscripten. These allow programs written in C/C++ to transparently perform common tasks such as using the file system, allocating memory and printing to the console. Emscripten allows a virtual file system to be created using its FS library, which is used by Emscripten's libc and libcxx for file I/O.[18] Files can be added or removed from the virtual file system using JavaScript helper functions. It is also possible to directly call C functions from JavaScript using Emscripten. These functions must first be named at compile time so they are not optimised out of the resulting compiled JavaScript code. The required functions are then wrapped using Emscripten's *cwrap* function and assigned to a JavaScript function name. The *cwrap* function allows many JavaScript variables to be used transparently as arguments to C functions, such as passing JavaScript strings to functions which require the C language *const char* array type.

Although Emscripten can successfully compile a large section of C/C++ code, there are still a number of limitations to this approach due to limitations within the JavaScript language and runtime. As JavaScript doesn't support threading, Emscripten is unable to compile codebases that make use of threads. Some concurrency is possible using web workers, but they do not share state. It is also not possible to directly implement 64-bit integers in JavaScript as all numbers are represented using 64-bit doubles. This results in a risk of rounding errors being introduced to the compiled JavaScript when performing arithmetic operations with 64-bit integers [14].

6.4.1 CsoundEmscripten

CsoundEmscripten is an implementation of the Csound language in JavaScript using the Emscripten compiler.[19]

---

[17]http://asmjs.org/spec/latest/.

[18]http://github.com/kripken/emscripten/wiki/Filesystem-API.

[19]A working example of CsoundEmscripten can be found at http://eddyc.github.io/CsoundEmscripten. The compiled Csound library and CsoundObj JavaScript class can be found at http://github.com/eddyc/CsoundEmscripten/.

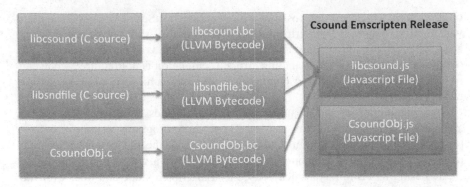

Fig. 6.1 The components of CsoundEmscripten

CsoundEmscripten consists of three main modules (see Fig. 6.1):

- The Csound library (and its dependency libsndfile) compiled to JavaScript using Emscripten
- A structure and associated functions written in C named CsoundObj implemented on top of the Csound library that is compiled to JavaScript using Emscripten
- A handwritten JavaScript class also named CsoundObj that contains the public interface to CsoundEmscripten. The JavaScript class both wraps the compiled CsoundObj structure and associated functions and connects the Csound library to the Web Audio API.

6.4.1.1   Wrapping the Csound C API for Use with JavaScript

In order to simplify the interface between the Csound C API and the JavaScript class containing the CsoundEmscripten public interface, a structure named CsoundObj and a number of functions which use this structure were created. The structure contains a reference to the current instance of Csound, a reference to Csound's input and output buffer and Csound's 0dBFS value. Some of the functions that use this structure are:

- CsoundObj_new(): This function allocates and returns an instance of the CsoundObj structure. It also initialises an instance of Csound and disables Csound's default handling of sound I/O, allowing Csound's input and output buffers to be used directly.
- CsoundObj_compileCSD(self, filePath, samplerate, controlrate, buffersize): This function is used to compile CSD files; it takes as its arguments a pointer to the CsoundObj structure self, the address of a CSD file given by filePath, a specified sample rate given by samplerate, a specified control rate given by controlrate and a buffer

size given by `buffersize`. The CSD file at the given address is compiled using these arguments.

- `CsoundObj_process(self, inNumberFrames, inputBuffer, outputBuffer)`: This function copies audio samples to Csound's input buffer and copies samples from Csound's output buffer. It takes as its arguments: a pointer to the `CsoundObj` structure `self`, an integer `inNumberFrames` specifying the number of samples to be copied, a pointer to a buffer containing the input samples named `inputBuffer` and a pointer to a destination buffer to copy the output samples named `outputBuffer`.

Each of the other functions that use the `CsoundObj` structure simply wraps existing functions present in the Csound C API. The relevant functions are:

- `csoundGetKsmps(csound)`: This function takes as its argument a pointer to an instance of Csound and returns the number of specified audio frames per control sample.
- `csoundGetNchnls(csound)`: This function takes as its argument a pointer to an instance of Csound and returns the number of specified audio output channels.
- `csoundGetNchnlsInput(csound)`: This function takes as its argument a pointer to an instance of Csound and returns the number of specified audio input channels.
- `csoundStop(csound)`: This function takes as its argument a pointer to an instance of Csound and stops the current performance pass.
- `csoundReset(csound)`: This function takes as its argument a pointer to an instance of Csound and resets its internal memory and state in preparation for a new performance.
- `csoundSetControlChannel(csound, name, val)`: This function takes as its arguments a pointer to an instance of Csound, a string given by `name` and number given by `val`; it sets the numerical value of a Csound control channel specified by the string `name`.

The `CsoundObj` structure and associated functions are compiled to JavaScript using Emscripten and added to the compiled Csound JavaScript library. Although this is not necessary, keeping the compiled `CsoundObj` structure and functions in the same file as the Csound library makes it more convenient when including CsoundEmscripten within web pages.

6.4.1.2   The CsoundEmscripten JavaScript Interface

The last component of CsoundEmscripten is the `CsoundObj` JavaScript class. This class provides the public interface for interacting with the compiled Csound library. As well as allocating an instance of Csound, this class provides methods for controlling performance and setting the values of Csound's control channels. Additionally, this class interfaces with the Web Audio API, providing Csound with samples from the audio input bus and copying samples from Csound to the audio output bus. Audio I/O and the Csound process are performed in JavaScript using the Web Audio API's `ScriptProcessorNode`. This node allows direct access to input and output samples in JavaScript allowing audio processing and synthesis using the Csound library.

Csound can be used in any web page by creating an instance of `CsoundObj` and calling the available public methods in JavaScript. The methods available in the `CsoundObj` class are:

- `compileCSD(fileName)`: This method takes as its argument the address of a CSD file `fileName` and compiles it for performance. The CSD file must be present in Emscripten's virtual file system. It calls the compiled C function `CsoundObj_compileCSD`. It also creates a `ScriptProcessorNode` instance for Audio I/O.
- `compileOrc(orc)`: This method takes a Csound orchestra `orc` as a string and compiles it for performance. It calls the compiled C function `CsoundObj_compileOrc`. Like the previous method, it also creates a `ScriptProcessor Node` instance for Audio I/O.
- `enableAudioInput()`: This method enables audio input to the web browser. When called, it triggers a permissions dialogue in the host web browser requesting permission to allow audio input. If permission is granted, audio input is available for the running Csound instance.
- `startAudioCallback()`: This method connects the `ScriptProcessor Node` to the audio output and, if required, the audio input. The audio processing callback is also started. During each callback, if required, audio samples from the `ScriptProcessorNodes` input are copied into Csound's input buffer and any new values for Csound's software channels are set. Csound's `csoundPerformKsmps()` function is called and any output samples are copied into the `ScriptProcessorNodes` output buffer.
- `stopAudioCallback()`: This method disconnects the current running `ScriptProcessorNode` and stops the audio process callback. If required this method also disconnects any audio inputs.
- `addControlChannel(name, initialValue)`: This method adds an object to a JavaScript array that is used to update Csound's named channel values. Each object contains a string value given by `name`, a float value given by `initialValue` and additionally a boolean value indicating whether the float value has been updated.
- `setControlChannelValue(name, value)`: This method sets a named control channel given by the string `name` to the specified number given by the `value` argument.
- `getControlChannelValue(name)`: This method returns the current value of a named control channel given by the string `name`.

6.4.1.3   A Simple Example

The following code for a HTML page shows an simple example of CsoundEmscripten. It runs a recursive instrument in Csound that calls itself creating a sequence
of sine wave pitches. Audio starts as the body of the page is loaded and the
successive Csound messages are displayed in the browser page.

```
<!DOCTYPE html>
<html>
<!--
 Csound Emscripten minimal example
 Copyright (C) 2014 V Lazzarini
-->
<head>
 <title>Minimal Csound Example</title>
</head>
<script src="javascripts/libcsound.js"></script>
<script src="javascripts/CsoundObj.js"></script>
<script type="text/javascript">

var csoundObj = new CsoundObj();

function onload(){

 // grab the console to display Csound messages
 console.log = function(message){
   var messField = document.getElementById("console")
   messField.innerText = message;
 }

 // compile the instrument
 csoundObj.compileOrc(
   "ksmps=256\n" +
   "0dbfs=1\n" +
   "instr 1 \n" +
   "k1 linen p4,0.1,p3,0.1 \n" +
   "a1 oscili k1,p5 \n" +
   "outs a1,a1 \n" +
   "schedule 1,0.25,0.5,0.1+rnd(0.1),500+rnd(500)\n" +
   "endin \n" +
   "schedule 1,0,0.5,0.1,500\n"
 );

 // start audio playback
 csoundObj.startAudioCallback();
```

```
}
</script>
<body onload="onload()">
  <div id="console"></div>
</body>
</html>
```

6.4.1.4 Limitations

Using CsoundEmscripten, it is possible to add Csound's audio processing and synthesis capabilities to any web browser that supports the Web Audio API. Unfortunately this approach of bringing Csound to the web comes with a number of drawbacks.

Although JavaScript engines are constantly improving in speed and efficiency, running Csound entirely in JavaScript is a processor-intensive task even on modern systems. This is especially troublesome when trying to run even moderately complex CSD files on mobile computing devices.

Another limitation is due to the design of the ScriptProcessorNode part of the Web Audio API. Unfortunately, the ScriptProcessorNode runs on the main thread. This can result in audio glitching when another process on the main thread—such as the UI—causes a delay in audio processing. As part of the W3C Web Audio Spec. review, it has been suggested that the ScriptProcessor Node be moved off of the main thread. There has also been a resolution by the Web Audio API developers that they will make it possible to use the ScriptProcessorNode with web workers.[20] Hopefully in a future version of the Web Audio API, the ScriptProcessorNode will be more capable of running the kind complex audio processing and synthesis capabilities allowed by the Csound library.

This version of Csound also does not support plugins, making some opcodes unavailable. Additionally, MIDI I/O is not currently supported. This is not due to the technical limitations of Emscripten; rather it was not implemented due to the current lack of support for the Web MIDI standard in Mozilla Firefox[21] and in the Webkit library.[22]

---

[20]http://www.w3.org/Bugs/Public/show_bug.cgi?id=17415#c94.

[21]http://bugzilla.mozilla.org/show_bug.cgi?id=836897.

[22]http://bugs.webkit.org/show_bug.cgi?id=107250.

6.5 Beyond Web Audio: Creating Audio Applications with PNaCl

As an alternative to the development of audio applications for web deployment in pure JavaScript, it is possible to take advantage of the Native Clients (NaCl) platform.[23] This allows the use of C and C++ code to create components that are accessible to client-side JavaScript and run natively inside the browser. NaCl is described as a sandboxing technology, as it provides a safe environment for code to be executed in an OS-independent manner [10, 12]. This is not completely unlike the use of Java with the Java Web start Technology, which has been discussed elsewhere in relation to Csound [8].

There are two basic toolchains in NaCl: native/gcc and Portable NaCl (PNaCl) [4]. While the former produces architecture-dependent code (arm, x86, etc.), the latter is completely independent of any existing architecture. NaCl is currently only supported by the Chrome and Chromium browsers. Since version 31, Chrome enables PNaCl by default, allowing applications created with that technology to work completely out of the box. While PNaCl modules can be served from anywhere in the open web, native-toolchain NaCl applications and extensions can only be installed from Google's Chrome Web Store.

6.5.1 The Pepper Plugin API

An integral part of NaCl is the Pepper Plugin API (PPAPI, or just Pepper). It offers various services, of which interfacing with JavaScript and accessing the audio device are particularly relevant to our ends. All of the toolchains also include support for parts of the standard C library (e.g. stdio), and very importantly for Csound, the pthread library. However, absent from the PNaCl toolchain are dlopen() and friends, which means no dynamic loading is available there.

JavaScript client-side code is responsible for requesting the loading of an NaCl module. Once the module is loaded, execution is controlled through JavaScript event listeners and message passing. A `PostMessage()` method is used by Pepper to allow communication from JavaScript to PNaCl module, triggering a message handler in the C/C++ side. In the opposite direction, a *message* event is issued when C/C++ code calls the equivalent `PostMessage()` function.

Audio output is well supported in Pepper with a mid-latency callback mechanism (ca. 10–11 ms, 512 frames at 44.1 or 48 kHz sampling rate). Its performance appears to be very uniform across the various platforms. The Audio API design is very straightforward, although the library is a little rigid in terms of parameters. It supports only stereo at one of the two sampling rates mentioned above. Audio input

---

[23]http://developers.google.com/native-client.

is not yet available in the production release, but support can already be seen in the development repository.

The most complex part of NaCl is access to the local files. In short, there is no open access to the client disk, only to sandboxed file systems. It is possible to mount a server file system (through httpfs), a memory file system (memfs) as well as local temporary or permanent file systems (html5fs). For those to be useful, they can only be mounted and accessed through the NaCl module, which means that any copying of data from the user disk into these partitions has to be mediated by code written in the NaCl module. For instance, it is possible to take advantage of the file HTML 5 tag and to get data from NaCl into a JavaScript blob so that it can be saved into the user's disk. It is also possible to copy a file from disk into the sandbox using the URLReader service supplied by Pepper.

6.5.2 PNaCl

The PNaCl toolchain compiles code down to a portable bitcode executable (called a *pexe*). When this is delivered to the browser, an ahead-of-time compiler is used to translate the code into native form. A web application using PNaCl will contain three basic components: the pexe binary, a manifest file describing it and a client-side script in JS, which loads and allows interaction with the module via the Pepper messaging system.

6.5.3 Csound for PNaCl

A fully functional implementation of Csound for Portable Native Clients is available as part of the Csound package releases.[24] The package is composed of three elements: the JavaScript module (csound.js), the manifest file (csound.nmf) and the pexe binary (csound.pexe). The source for the PNaCl component is also available from that site (csound.cpp). It depends on the Csound and libsndfile libraries compiled for PNaCl and the NaCL sdk (see Fig. 6.2). A makefile for PNaCl exists in the Csound 6 sources.

Users of Csound for PNaCl will only interact with the services offered by the JavaScript module. Typically an application written in HTML 5 will require the following elements to use it:

1. The csound.js script
2. A reference to the module using a div tag with id="engine"
3. A script containing the code to control Csound

---

[24]http://sourceforge.net/projects/csound/files/csound6. This is also available at http://vlazzarini.github.io, together with a series of examples and reference documentation.

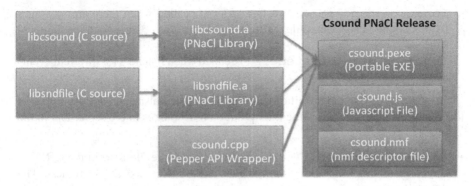

Fig. 6.2 The components of Csound for PNaCl

6.5.3.1   The JavaScript Interface

A comprehensive set of methods are available in csound.js to control Csound:

- `csound.Play()`: starts performance.
- `csound.PlayCsd(s)`: starts performance from a CSD file s, which can be in ./http/ (ORIGIN server) or ./local/ (local sandbox).
- `csound.RenderCsd(s)`: renders a CSD file s, which can be in ./http/ (ORIGIN server) or ./local/ (local sandbox), with no RT audio output. The "finished render" message is issued on completion.
- `csound.Pause()`: pauses performance.
- `csound.CompileOrc(s)`: compiles the Csound code in the string s.
- `csound.ReadScore(s)`: reads the score in the string s (with preprocessing support).
- `csound.Event(s)`: sends in the line events contained in the string s (no preprocessing).
- `csound.SetChannel(name, value)`: sets the control channel name to the value value, both arguments being strings.
- `csound.RequestChannel(name)` –requests the value of the output control channel name. The channel value will be passed from Csound as a message in the format ::control::channel:value.
- `csound.RequestTable(num)`: requests the data from a function table in Csound.
- `csound.GetTableData()`: returns the most recently requested table data.
- `SetStringChannel(name, string)`: sets the string channel name to the string string, both arguments being strings.
- `StartInputAudio()`: starts audio input into Csound.

As it starts, the PNaCl module will call a `moduleDidLoad()` function, if it exists. This can be defined in the application script. Also the following callbacks are also definable:

- `function handleMessage(message)`: called when there are messages from Csound (PNaCl module). The string `message.data` contains the message.
- `function attachListeners()`: called when listeners for different events are to be attached.

In addition to Csound-specific controls, the module also includes a number of file system facilities, to allow the manipulation of resources in the server and in the sandbox:

- `csound.CopyToLocal(src, dest)`: copies the file `src` in the ORIGIN directory to the local file `dest`, which can be accessed at ./local/dest. The "Complete" message is issued on completion.
- `csound.CopyUrlToLocal(url,dest)`: copies the url `url` to the local file `dest`, which can be accessed at ./local/dest. Currently only ORIGIN and CORS urls are allowed remotely, but local files can also be passed if encoded as urls with the webkitURL.createObjectURL() JavaScript method. The "Complete" message is issued on completion.
- `csound.RequestFileFromLocal(src)`: requests the data from the local file `src`. The "Complete" message is issued on completion.
- `csound.GetFileData()`: returns the most recently requested file data as an ArrayObject.

6.5.3.2 An Introductory Example

A minimal example of the Csound PNaCl functionality is shown below. After the module is loaded, Csound is started and we send in a simple instrument containing a sine wave ping with random pitch within a given range (one octave above A4). We attach an event handler to listen for mouse clicks on the page. This will trigger a note on the instrument we loaded, and Csound will print information messages to the page (Fig. 6.3).

```
<!DOCTYPE html>
<html>
<!--
 Csound pnacl minimal example
 Copyright (C) 2013 V Lazzarini
-->
<head>
 <title>Minimal Csound Example</title>
 <script type="text/Javascript" src="csound.js">
 </script>
 <script type="text/javascript">

  // called by csound.js on load
  function moduleDidLoad() {
```

```
      // start Csound
      csound.Play();
      // compile a simple instrument
      csound.CompileOrc(
         "instr 1 \n" +
         "k1 expon 1,p3,0.001 \n" +
         "a1 oscili 0.1*k1,440+rnd(440) \n" +
         "outs a1,a1 \n" +
         "endin");
      // tell the user what to do
      document.getElementById("tit").innerHTML
         = "Click on the page below to play";
   }
   // called by csound.js to attach event handler
   function attachListeners() {
      document.getElementById("mess").
            addEventListener("click",Play);
   }

   // default event handling function called by csound.js
   function handleMessage(message) {
      var mess = message.data;
      var messField = document.getElementById("mess")
      messField.innerText += mess;
   }

   // click handler, called when user clicks on page
   function Play() {
      csound.Event("i 1 0 5");
   }
  </script>
</head>
<body>
  <div id="console"></div>
   <h3 id="tit"> </h3>
  <div id="mess">

  </div>
  <!--pNaCl csound module-->
  <div id="engine"></div>

</body>
</html>
```

Csound: loading... (count=1)
Csound: ready

Click on the page below to play

Csound: running...
sample rate overrides: esr = 44100.0000, ekr = 689.0625, ksmps = 64
--Csound version 6.03.1 (float samples) May 7 2014
graphics suppressed, ascii substituted
0dBFS level = 1.0
orch now loaded
audio buffered in 512 sample-frame blocks
SECTION 1:
rtevent: T 11.889 TT 11.889 M: 0.00000 0.00000
new alloc for instr 1:
rtevent: T 12.562 TT 12.562 M: 0.09999 0.09999
new alloc for instr 1:
rtevent: T 13.375 TT 13.375 M: 0.13888 0.13888
new alloc for instr 1:
rtevent: T 17.438 TT 17.438 M: 0.13980 0.13980
rtevent: T 18.437 TT 18.437 M: 0.10022 0.10022
rtevent: T 24.683 TT 24.683 M: 0.12424 0.12424
rtevent: T 25.496 TT 25.496 M: 0.09998 0.09998
rtevent: T 27.028 TT 27.028 M: 0.13102 0.13102
rtevent: T 30.337 TT 30.337 M: 0.10803 0.10803

Fig. 6.3 A simple Csound for PNaCl example

In addition to this one, a series of more complex examples demonstrating this API is provided in the Csound main repository at GitHub.[25]

6.5.3.3 Limitations

The following limitations apply to the current release of Csound for PNaCl:

- No MIDI in the NaCl module. However, it might be possible to implement MIDI in JavaScript (through Web MIDI) and, using the csound.js functions, send control data to Csound and respond to the various channel messages.

---

[25]http://www.github.com/csound.

- No plugins, as PNaCl does not support dlopen() and friends. This means some Csound opcodes are not available as they reside in plugin libraries. It might be possible to add some of these opcodes statically to the Csound PNaCl library in the future.

Conclusions

In this chapter we reviewed the current state of support for the development of web-based audio applications for Ubiquitous Music. As part of this, we explored two approaches in deploying Csound as an engine for general-purpose media software. The first consisted of a JavaScript version created with the help of the Emscripten compiler and the second a native C/C++ port for the Native Client platform, using the Portable Native Client toolchain. The first has the advantage of enjoying widespread support by a variety of browsers but is not yet fully deployable. On the other hand, the second approach, while at the moment only running on Chrome and Chromium browsers, is a robust and ready-for-production version of Csound.

Acknowledgements This research was partly funded by the Program of Research in Third-Level Institutions (PRTLI 5) of the Higher Education Authority (HEA) of Ireland, through the Digital Arts and Humanities programme.

References

1. Batchelor, P., Wignall, T.: BeaglePi: an introductory guide to Csound on the BeagleBone and the Raspberry Pi, as well other Linux-powered tinyware. Csound J. **18** (2013). www.csounds.com/journal/issue18/beagle_pi.html
2. Boulanger, R.J. (ed.): The Csound Book: Tutorials in Software Synthesis and Sound Design. MIT Press, Cambridge (2000)
3. Casey, M., Smaragdis, P.: Netsound. In: On the Edge. ICMA and HKUST (1996)
4. Donovan, A., Muth, R., Chen, B., Sehr, D.: PNaCl: Portable Native Client Executables. Google White Paper (2010)
5. ffitch, J., Mitchell, J., Padget, J.: Composition with sound web services and workflows. In: Ltd S.O. (ed.) Proceedings of the 2007 International Computer Music Conference, vol. I, pp. 419–422. ICMA and Re:New (2007). ISBN 0-9713192-5-1
6. Johannes, T., Toshihiro, K.: " 'Và, pensiero!' " - Fly, thought! Experiment for interactive internet based piece using Csound6 (2013). http://tarmo.uuu.ee/varia/failid/cs/pensiero-files/pensiero-presentation.pdf. Accessed 2 Feb 2014
7. Lazzarini, V., Yi, S., Timoney, J.: Digital audio effects on mobile platforms. In: Proceedings of DAFx 2012 (2012)
8. Lazzarini, V., Yi, S., Timoney, J., Keller, D., Pimenta, M.: The Mobile Csound Platform. In: Proceedings of ICMC 2012 (2012)
9. Roberts, C., Wakefield, G., Wright, M.: The Web Browser as Synthesizer and Interface. In: Proceedings of the International Conference on New Interfaces for Musical Expression (2013)

10. Sehr, D., Muth, R., BifiñĆ, Khimenko, V., Pasko, E., Schimpf, K., Yee, B., Chen, B.: Adapting software fault isolation to contemporary CPU architectures. In: 19th USENIX Security Symposium (2010)
11. Wyse, L., Subramanian, S.: The viability of the Web browser as a computer music platform. Comput. Music J. **37**(4), 10–23 (2013)
12. Yee, B., Sehr, D., Dardyk, G., Chen, J.B., Muth, R., Ormandy, T., Okasaka, S., Narula, N., Fullagar, N.: Native client: A Sandbox for portable, untrusted x86 native code. In: 2009 IEEE Symposium on Security and Privacy (2009)
13. Yi, S., Lazzarini, V.: Csound for Android. In: Linux Audio Conference, vol. 6 (2012)
14. Zakai, A.: Emscripten: an llvm-to-javascript compiler. In: Proceedings of the ACM International Conference Companion on Object Oriented Programming Systems Languages and Applications, pp. 301–312. ACM, New York (2011)

Chapter 7
Ubiquitous Music Ecosystems: Faust Programs in Csound

Victor Lazzarini, Damián Keller, Marcelo Pimenta, and Joseph Timoney

Abstract This chapter describes the combination of two high-level audio and music programming systems, Faust and Csound. The latter is a MUSIC N-derived language, with a large set of unit generators and a long history of development. The former is a purely functional language designed to describe audio processing algorithms that can be compiled into a variety of formats. The two systems are combined in the Faust Csound opcodes, which allow the on-the-fly programming, compilation and instantiation of Faust DSP programs in a running Csound environment. Examples are presented, and the concept of Ubiquitous Music Ecosystem is discussed.

7.1 Introduction

Audio signal processing algorithms can be expressed and implemented in a variety of environments. These range from the lower level of microcode and assembler programming to high-level matrix-manipulation programs such as MatLab and Octave and patching systems such as PD or MaxMSP [11]. In general, the advantage of higher-level specifications is that the algorithm is presented compactly in encapsulating blocks, which afford good readability and are easily manipulated. On the other hand, such gains are normally accompanied by a loss of computational

V. Lazzarini (✉)
Music Department, Maynooth University, Ireland
e-mail: vlazzarini@nuim.ie

D. Keller
Amazon Center for Music Research - NAP, Federal University of Acre, Rio Branco, Brazil
e-mail: dkeller@ccrma.stanford.edu

M. Pimenta
Institute of Informatics, UFRGS, Universidade Federal do Rio Grande do Sul, Porto Alegre, Brazil
e-mail: mpimenta@inf.ufrgs.br

J. Timoney
Computer Science Department, Maynooth University, Ireland
e-mail: jtimoney@nuim.ie

© Springer International Publishing Switzerland 2014   129
D. Keller et al. (eds.), *Ubiquitous Music*, Computational Music Science,
DOI 10.1007/978-3-319-11152-0__7

efficiency, especially in the case of general-purpose systems. In high-level real-time audio programming systems, where processes can be run efficiently, there is a limit of what can be expressed, if compared to lower-level environments.

In this scenario, we find that languages that can sit at a middle-level in terms of complexity are optimally placed to provide efficiency and generality to allow the design and implementation of audio processes. In this chapter, we will describe the combination of two such systems, Csound and Faust, in the development of support tools for Ubiquitous Music making [3]. The chapter is organised as follows. First, we will introduce the two systems and discuss their characteristics. The embedding of Faust in Csound will then be detailed, with some use examples. Finally, the chapter will conclude with a discussion of a proposal for a new concept for audio programming in a multi-language environment: the Ubiquitous Music Ecosystem.

7.2 Csound

Csound [12] is an heir to the MUSIC N systems derived from Mathews' MUSIC IV[8]. Although it still allows a traditional score + orchestra programming approach, it is not limited to it. The system is built around a library [2] that is accessed through its application programming interface (API) and is manipulated via a variety of front ends, the most basic of them being the command-line interface (CLI) program `csound`. The API can be used directly from a number of languages (C, C++, Java, Clojure, Python, Lua, among others).

7.2.1 Csound Programming

Most of Csound programming is done through its orchestra language. In this, the majority of the code is structured around blocks called *instruments*, where it is based on simple statements such as

```
[out,...] opcode [in, ...]
```

   or

```
[out =] opcode([in,...])
```

where `opcode` is a given unit generator that will either generate an output, process or just consume an input. In the second form, inputs can be taken directly from other opcodes, by function composition. In general, inputs can be expressions of any complexity.

Instruments are defined between the keywords `instr` and `endin`. For instance,

```
instr 1
  i1 = 1000
  a1 rand i1
     out a1
endin
```

defines such an instrument, which is given the number 1 as an identifier. With this defined, it is possible to instantiate it, so that it can perform its processing. The following line runs an instance of instrument 1 for 1 s starting at time 0 (now). A running instance of an instrument also known as an *event* or, more traditionally, a *note*.

```
schedule 1, 0, 1
```

In this case, when we compile the above code (both the instrument definition and the schedule statement), we will have a random number generator producing white noise. It is possible to schedule more events of the same instrument. There are no limits placed on the number of simultaneous instances of an instrument.

Instruments contain code that is executed sequentially in two separate stages: at initialisation time (init-time only once) and at performance time (perf-time, continuously in an implicit loop). In the example above, the first line, `i1 = 1000`, is executed at init-time only, whereas the second will be executed at init-time (where the opcode `rand` is initialised), and at perf-time, producing audio at its output. The third line, `out a1`, is executed at perf-time only. The Csound compiler automatically assigns code to init- or perf-time depending on the types of variables used, which are defined by the first letter of their name. For instance, init-time numeric variables start with `i`, whereas `a` types are audio variables and thus imply code that is run at perf-time.

When Csound runs, there is an implicit perf-time loop, which makes the instrument compute audio in blocks of `ksmps` samples, so that a-variables are constantly updated with new blocks of samples. There are other variable types, which follow similar rules. Particularly relevant to our discussion is the k-type, which holds a single sample at perf-time and is normally used to carry control signals.

Any code placed outside instruments (known as 'global' or 'instrument 0' code) can only use init-time operations, as the performance pass is exclusive to instruments. So in this case, when we schedule an event from a global statement, we do that only once, at the time the code is compiled. It is possible however to create loops to repeat the schedule operation a number of times programmatically, as Csound has a set of control-flow and looping constructs that can be used at both initialisation and performance stages. It is also possible to re-compile the schedule operation and launch another instance of instrument 1. The compiler can be used at any time to run global code or to add or replace instruments.

Recursion is another useful method to generate more than one event. If we place a schedule statement inside an instrument, it will be able schedule itself when it runs:

```
instr 1
 k1 linen   p4,0.01,p3,0.1
 a1 oscili k1, p5
     out a1
 schedule 1, 0.5, p3, p4, p5
endin
```

```
schedule 1, 0, 0.5, 1000, 440
```

In this case, we will have a new event every 0.5 s, which will be a sine wave ping at 440 Hz. Note that event parameters are referred to in an instrument by the letter "p" followed by the parameter number. The sequence set up by this code is never ending, but it is possible to stop it by compiling an empty instrument 1 to replace the one above.

Csound has a full complement of arithmetic operators, mathematical functions and control-flow structures, in addition to over 1800 unit generators (opcodes). It can be used to describe any time-domain audio signal processing algorithm. It has also special frequency-domain types and opcodes, which can be used to design spectral processing instruments, and arrays of all internal data types, single or multi-dimensional. All aspects of the Csound processing engine can be configured with system parameters passed to it at the start of a session.

Csound can take orchestra code from plain text files. Alternatively, it can also work unified Csound definition files (CSDs). These have an XML-like format, with tags defining sections of the code, and can include, in addition to the orchestra code, options to the system, a score (using the Csound score language) and base-64 encoded media files. The following is an example of CSD file that uses the orchestra code discussed above:

```
<CsoundSynthesizer>
<CsOptions>
 -odac
</CsOptions>
<CsInstruments>
 schedule 1, 0, 1
 instr 1
  i1 = 1000
  a1 rand i1
      out a1
 endin
</CsInstruments>
</CsoundSynthesizer>
```

7.2.2   The Csound API

The Csound system can be fully accessed via an API. With this, it is possible to control it programmatically from a variety of environments. The API is originally provided in the C language, but it has wrappers for a number of alternative languages: C++, Java, Python, Common Lisp, Clojure, Go, Lua, to cite but a few. In this article, we will concentrate on aspects of the API in Python 2.7.

The latest Csound module is accessed with the following command:

```
import csnd6
```

With the module loaded, we can create a Csound object. As the library is fully re-entrant, any number of these objects can be created, representing an instance of the whole system. The Csound class encapsulates all the operations to read, compile and run code written in the Csound language.

```
csound = csnd6.Csound()
```

To compile an orchestra such as the ones given in the examples of Sect. 7.2.1, we can invoke the CompileOrc method:

```
csound.CompileOrc('''
instr 1
 k1 linen  p4,0.01,p3,0.1
 a1 oscili k1, p5
    out a1
 schedule 1, 0.5, p3, p4, p5
endin

schedule 1, 0, 0.5, 1000, 44
''')
```

No sound will be heard until we start the Csound engine, set up the output to "dac" (the digital to analog converter, which means the computer soundcard) and run the performance loop. This can be done in the current thread or in a separate one. The latter option is generally more useful as we can use the main thread to control Csound as it runs. For this, we create a CsoundPerformanceThread object that will set up the background thread and manage it.

```
csound.SetOption("-odac")
csound.Start()
perf = csnd6.CsoundPerformanceThread(csound)
perf.Play()
```

At any point, we can call the compiler to add or replace instruments and to schedule them. It is also possible to launch new events using the Csound score language. In this case, instrument instances are defined in blank space-separated lists. These can be passed to Csound using the `ReadScore` method:

```
csound.ReadScore('''
i 1 0 1 1000 440
i 1 1 1 2000 660
''')
```

In this format, the "i" identifier indicates we want to start a new event, with the parameters following it (instrument, start, duration, p4, p5), separated by spaces.

In addition to the methods discussed above, the Csound class has extensive facilities for system setup, compilation from strings or files (for instance, from CSD files), MIDI and audio input/output configuration, software bus (for accessing audio, control and frequency-domain signals inside Csound), score handling, function table access and other miscellaneous operations.[1]

7.3 Faust

Faust [9] is a purely functional language designed to describe audio streams, with which we can implement any time-domain audio processing algorithm. Its compiler can produce C, C++, Javascript or LLVM code. The compiled code is an efficient audio digital signal processing (DSP) program that can be then used in a variety of environments (as plugins to various systems, including Csound, or as stand-alone programs). Signal processing programs created in Faust will generally be more efficient than their equivalent code written in other high-level music programming languages (Csound included).

7.3.1 Faust Programming

The Faust program describes a process. For instance, the following minimal program implements a mixer of two input signals:

```
process = + ;
```

Each Faust statement is terminated by a semicolon (`;`). So here we have an arithmetic operator (+), which by definition takes two inputs and produces one output. Summing two signals is the same as mixing them. Implicit in this operation is the fact that we will take two inputs and produce one output.

Similarly, if we want to scale a signal by 2, we can have

```
process = *(2);
```

---

[1] http://csound.github.io/docs/api/index.html.

which is a program of one input and one output, because the multiplication by 2 is a function of one input to one output. Again, the number of inputs and outputs is implicit.

Faust programs can also use the sequential (:) and parallel (,) operators:

```
process = _ ,2 : *;
```

This takes some audio input (_), in parallel with the constant 2, and sends them in sequence (:) to the function "*". Taken as whole, this process is equivalent to the previous example, as it takes one input and multiplies it by 2. The (_) is a placeholder, indicating that some input is expected there.

The other important primitives are split (<:) and merge (:>). The former turns one input into two (containing the same signal), and the other takes two signals and adds them up. With them, we can organise the flow of signals in a very convenient way.

For instance, this is how we square a signal:

```
process =  _ <: *;
```

Another version of the mix-two program above can be written as

```
process = _,_:> _;
```

Some specific signal processing operators are also defined, such as the "@" delay, which outputs a signal delayed by a certain number of samples. With it, we can do effects such as this slapback echo:

```
process = _<:_,_@4410:> _;
```

The input signal gets split into two streams, one of which is delayed by 4,410 samples (0.1 s at 44,100 samples per seconds) and mixed back together again. This is shown in Fig. 7.1, which was generated by the Faust compiler directly from this code.

For very short delays, the single quote and double quotes can be used to signify one- or two-sample delays. A simple first-order average can be written as

```
process = _<:_,_':>*(0.5);
```

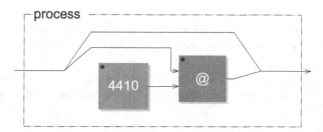

Fig. 7.1 A slapback echo

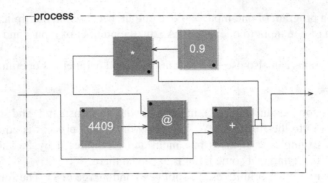

Fig. 7.2 Feedback delay

The language allows separate functions to be defined and applied. For instance, the previous example might be rewritten for clarity as a function of one input:

```
f(x) = (x + x')/2;
process = f;
```

Once we define the function, we can use it in our process. In this case, the process is nothing more than applying the function to an input. Finally, Faust also includes the concept of feedback, which has an implicit one-sample delay. This is defined by the operator "~" (tilde). It can be used to feed the output of an operation back into its input. For instance, a feedback delay can be constructed with this operator, using the basic slapback echo shown above:

```
process =    @(4409) +_ ~ *(0.9);
```

To the left of the tilde operator, we have an expression of two inputs, the signal and the feedback. The feedback signal is delayed by a further 4,409 samples (making the total delay 4,410 samples) and also scaled by 0.9. The output of the whole function is a sum of these two inputs (Fig. 7.2).

The recursion provided by the tilde operator can also be used to create sequences in the absence of loop constructions. For instance, a sample counting expression can be defined by

```
time = +(1) ~ _ ;
```

This operation combines a signal identity (_), which is the output, unmodified, and the function +(1), which adds one to it, recursively. The result is the sequence 1,2,3,....

Faust also allows access to C library functions for trigonometric operations, etc. With these and the time-counting function time, we can, for instance, write a sine wave oscillator as

```
process = time : *(2*PI/SR) : *(440) : sin;
```

where PI and SR are constants set to π and the sampling rate.

Faust includes a number of generic user interface (UI) functions that can be used to create controls. These get compiled to various forms, depending on the target of the compilation. For instance, a horizontal slider is defined by the following line:

```
freq = hslider("frequency", 440, 100, 1000, 1)
```

where the parameters are, respectively, name label, default value, minimum, maximum and minimum step. UI functions are implemented in different ways, depending on the system that hosts the Faust program. For instance, in a graphical user interface (GUI) program, a `hslider` will be an actual horizontal slider. In Csound, which does not define a specific GUI, this will be translated as a generic control-rate parameter.

7.3.2 Architectures

Faust programs can be compiled into C code for a variety of *architectures*, which are defined for several target systems. For instance, it is possible to generate code for plugins for various systems: Steinberg's Virtual Studio Technology (VST), Adobe Flash, MaxMSP, LADSPA, Supercollider, Pure Data, SndRT and Csound. In addition, the Faust compiler can create stand-alone programs using Coreaudio, Jack Connection Kit, Alsa, Open Sound System or Portaudio backends, with QT or GTK GUI.

The plugins or stand-alone programs are built from the generated source code using standard C/C++ compiler tools and libraries. In the case of Csound, the plugins are dynamic-link modules that are loaded into the system on start-up. The Faust program is then used in Csound as another opcode in the system. It is not only a viable alternative to writing new components directly in C or C++, but it also allows the programmer to concentrate on the specific aspects of the signal processing algorithm.

Although this system works well, it involves an extra step that requires a set of development tools and a certain expertise to complete it. This step also keeps the target system, in this case Csound, separate from the opcode implementation, in the Faust language. Any modifications to the code require the re-building and reloading of the dynamic library that contains it.

7.3.3 The Faust Library

With Faust version 2, however, the compiler system has been redesigned into a library, libfaust. This allows the embedding of the compiler into other programs and environments. In this scenario, Faust produces LLVM [4] bitcode via a just-in-time compiler and provides means of executing it. This allows the complete compilation and running of a Faust program to happen on the fly, under a host, which in this case is Csound.

There are three basic steps in the Faust dynamic compilation and performance:

1. Compilation: Faust code is compiled into a DSP factory object. This is a binary representation of the processing code that can be instantiated.
2. Instantiation: From a DSP factory object, we can create a process instance, which is ready to be run and performs the desired signal processing.
3. Performance: With an instance in memory, it is possible to invoke its compute() to produce the audio output.

The following functions are used to compile Faust code:

```
llvm_dsp_factory*
createDSPFactoryFromFile(const std::string& filename,
     int argc, const char *argv[],
     const std::string& target,
     std::string& error_msg, int opt_level = 3);

llvm_dsp_factory*
createDSPFactoryFromString(const std::string& name_app,
        const std::string& dsp_content,
        int argc, const char *argv[],
        const std::string& target,
        std::string& error_msg, int opt_level = 3);
```

They translate the Faust code into a LLVM bitcode form which contains the blueprint of a Faust process. This now resides memory and can be used in a program to instantiate a DSP object that can be executed. The following function is responsible for this:

```
llvm_dsp* createDSPInstance(llvm_dsp_factory* factory);
```

The llvm_dsp class which encapsulates the process has the following public methods:

```
class llvm_dsp : public dsp {
    public:
        virtual int getNumInputs();
        virtual int getNumOutputs();
        virtual void init(int samplingFreq);
        virtual void buildUserInterface(UI* inter);
        virtual void compute(int count,
            FAUSTFLOAT** input, FAUSTFLOAT** output);
};
```

With these, we can obtain the number of inputs and outputs, supply a user interface to control the process, as well as initialise and perform the process. The following code fragment demonstrates how a DSP object is used:

```
class controls : public UI { ...};
...
dsp = createDSPInstance(factory);
ctls = new controls;
dsp->buildUserInterface(ctls);
dsp->Init(sr);

for(n=0; n < end; n +=blocksize) {
...
dsp->compute(blocksize, input, output);
...
}
```

After the DSP object is created, we can attach a user interface to it. User interfaces are defined by objects that inherit from the UI base class supplied by the Faust library. The initialisation method is then invoked by passing the current sampling rate value to it. To process audio, we execute the DSP object by calling its compute() method, which consumes a block of input samples, and produces a block of output samples. The size of the block can be variable, and is passed at every call to this method.

7.4   The Faust Csound Unit Generators

The Faust Csound unit generators, or opcodes, have been built to take advantage of the capabilities offered by this library. They can take an arbitrary Faust program, compile it and then create running instances of it within a Csound instrument. So it is possible to implement a completely new process, from first principles and run it as efficiently as compiled C code.

The design of these opcodes mirrors the facilities discussed in the previous section. In order to simplify their use within the Csound system, the compilation and performance steps are made separate. This follows the fact that initialisation and performance in Csound are placed into two distinct phases. First, all opcodes that have initialisation tasks to execute are run once and only once as the instrument that contains them is instantiated. Some opcodes will only be run at this stage, as they do not process signal. At performance time, we then have code that is invoked repeatedly in a loop to process audio.

The compilation step in Faust only needs to be run at initialisation time. It does not involve any signal processing, and it does need to be repeated. Separating it from a DSP performance opcode also allows any number of DSP objects to be created from the same factory, which is a common application scenario.

The performance aspect of the Faust integration is subdivided into two elements, signal processing and controls, and exists in two separate opcodes. The signal processing part is only concerned in picking up any input audio (if it exists), doing something it and passing it to the output. In order to do parameter control, we use another opcode that can adjust a given user interface element defined in the DSP.

Finally, there is also an opcode that has been designed for single-instance DSPs, which integrates a compilation phase at init-time and signal processing at perf-time. This opcode can be employed for 'one-off' effects that are not designed to be run in multiple instances.

7.4.1 Faustcompile

The `faustcompile` opcode invokes the just-in-time compiler to produce a instantiable DSP process from a Faust program. It will compile a Faust program from a string, controlled by various arguments. Multiline strings are accepted, using {{ }} to enclose the string.

7.4.1.1 Syntax

```
ihandle faustcompile Scode, Sargs
```

  Scode: a string (in double quotes or enclosed by {{ }}) containing a Faust program
  Sargs: a string (in double quotes or enclosed by {{ }}) containing Faust compiler args

7.4.1.2 Example

```
ihandle faustcompile "process=+;", "-vec -lv 1"
```

7.4.2 Faustaudio

The `faustaudio` opcode instantiates and runs a compiled Faust program. It will work with a program compiled with `faustcompile`.

7.4.2.1   Syntax

```
ihandle,a1[,a2,...] faustaudio ifac[,ain1,...]
```

ifac: a handle to a compiled Faust program, produced by faustcompile
ihandle: a handle to the Faust DSP instance, which can be used to access its
controls with faustctl
ain1,...: input signals
a1,...: output signals

7.4.2.2   Example

```
ifac faustcompile "process=+;", "-vec -lv 1"
idsp,a1 faustaudio ifac,ain1,ain2
```

7.4.3   Faustctl

The faustctl opcode adjusts UI controls in a Faust DSP instance. It will set a
given control in a running faust program.

7.4.3.1   Syntax

```
faustctl idsp,Scontrol,kval
```

Scontrol: a string containing the control name
dsp: a handle to an existing Faust DSP instance
kval: value to which the control will be set

7.4.3.2   Example

```
idsp,a1 faustgen
gain = hslider("vol",1,0,1,0.01);
process = (_ * gain);
, ain1
faustctl idsp, "vol", 0.5
```

7.4.4 Faustgen

The opcode faustgen compiles, instantiates and runs a compiled Faust program.
It will invoke the just-in-time compiler at i-time and instantiate the DSP program.
At perf-time, it will run the compiled Faust code.

7.4.4.1 Syntax

```
ihandle,a1[,a2,...] faustgen SCode[,ain1,...]
```

Scode: a string containing a Faust program
  ihandle: a handle to the Faust DSP instance, which can be used to access its
controls with faustctl
ain1,...: input signals
a1,...: output signals

7.4.4.2 Example

```
idsp,a1 faustgen "process=+;",ain1,ain2
```

7.5 Examples

In this section, we explore three examples of Faust programs used to implement
opcodes in Csound. The first example illustrates some aspects of Faust programming
discussed above in Sect. 7.3. The second shows a classic algorithm from the
Faust library used within a Csound instrument for MIDI-based performances. The
final example demonstrates Faust programs for sound transformation and how a
new algorithm from the literature can be brought directly into computer music
performance.

7.5.1 A Sine Wave Generator

Our first example shows the complete sine wave oscillator example discussed in
Sect. 7.3 (Fig. 7.3). The program is placed in one-off faustgen opcode (the {{ and
}} enclose a multiline string in Csound), and faustctl is used to set the oscillator
frequency:

```
instr 1

  idsp, asig faustgen {{
  PI  = 3.1415926535897932385;
```

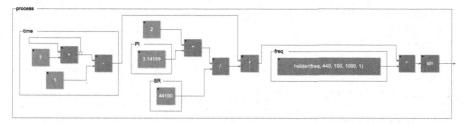

Fig. 7.3 Sine wave oscillator

Fig. 7.4 Karplus–Strong algorithm

```
SR = 44100;
freq = hslider("freq", 440,100,1000,1);
time = (+(1) ~ _ ) - 1;
process =    time : *(2*PI/SR) : *(freq) : sin;
}}

faustctl idsp,"freq", 440

   out asig
endin
```

7.5.2   Karplus–Strong Synthesizer

The second example is a bit more involved, showing the classic Karplus–Strong program in Faust (Fig. 7.4), embedded in a Csound orchestra that can be controlled via MIDI (in a complete CSD source code). The program for this example comes from the Faust instrument library, and it uses functions and definitions from the file "music.lib", that is distributed with the system (note the import function being used for this purpose). The faustcompile opcode is placed at global level in Csound, so it is run only once, but its factory can be instantiated by any instruments in the orchestra:

```
<CsoundSynthesizer>
<CsOptions>
--midi-velocity-amp=4 --midi-key-cps=5
</CsOptions>
```

```
<CsInstruments>
ksmps=100
nchnls=2
0dbfs = 1

giPluck faustcompile {{

import("music.lib");
upfront(x)        = (x-x') > 0.0;
decay(n,x)        = x - (x>0.0)/n;
release(n)        = + ~ decay(n);
trigger(n)        = upfront : release(n) : >(0.0);

size = hslider("excitation", 128, 2, 1024, 1);
dur = hslider("duration", 128, 2, 1024, 1);
att  = hslider("attenuation", 0.1, 0, 1, 0.01);

average(x) = (x+x')/2;
resonator(d, a) = (+ : delay(4096, d-1.5))
                              ~ (average : *(1.0-a)) ;
process = noise * hslider("level", 0.5, 0, 1, 0.01)
: vgroup("excitator", *(button("play"): trigger(size)))
: vgroup("resonator", resonator(dur, att));

}}, "-vec -lv 1"

instr 1
 i3, a1 faustaudio giPluck
 faustctl i3,"level", p4*0.5
 faustctl i3,"duration", sr/(p5)
 faustctl i3,"excitation", sr/(p5)
 faustctl i3,"attenuation", 0.01
 faustctl i3,"play", 1
 kenv linsegr 1,1,1, 0.01, 0
   outs a1*kenv,a1*kenv
endin

</CsInstruments>
</CsoundSynthesizer>
```

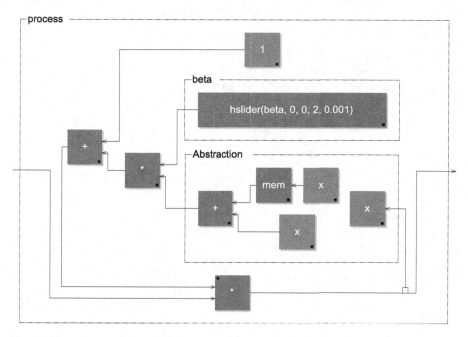

Fig. 7.5 Second-order feedback AM

7.5.3 Effects

In this example, we demonstrate how Faust opcodes can be used to process audio. We have two opcodes: a second-order feedback amplitude modulation (Fig. 7.5) [5] processor that distorts a sine wave to create a signal rich in harmonics, which is used in instrument 1. The generated sounds are then processed by a global effects instrument containing the second Faust program, a delay effect based on a comb filter. This last process is run inside a faustgen opcode, which is useful in cases where only a single instance of the Faust program is required.

This example also demonstrates how a novel algorithm (second-order feedback AM) can be easily brought into a fully functional system for computer music.

```
<CsoundSynthesizer>
<CsOptions>
--midi-velocity-amp=4 --midi-key-cps=5 -d -odac
</CsOptions>
<CsInstruments>
0dbfs = 1
nchnls= 2

ga1 init 0
```

```
schedule 101,0,-1

gifbam2 faustcompile {{
beta = hslider("beta", 0, 0, 2, 0.001);  .
fbam2(b) = *~(((\(x).(x + x'))*b)+1);
process = fbam2(beta);
}}, "-vec -lv 1"

instr 1

ain oscili 1, p5, -1, 0.25

ival = p4*2
if ival > 1.2 then
ival = 1.2
endif
if ival < 0.5 then
ival = 0.5
endif

kb1 expsegr ival,0.01,ival,27,0.001,0.2,0.001

ib, asig faustaudio gifbam2,ain
faustctl ib,"beta",kb1

asig balance asig, ain

kenv expsegr 1,5,0.001,0.2,0.001
aenv2 linsegr 0,0.015,0,0.001,p4,0.2,p4

aout = asig*aenv2*kenv*0.5
ga1= ga1 + aout*0.2
outs  aout,aout
endin

instr 101
iCmb,a1 faustgen {{
sr = 44100;
delay = hslider("delay", 0, 0, 1, 0.001);
comb(t,g) = + @(t*sr) ~ *(g);
process = comb(delay,0.5);
}}, ga1
faustctl iCmb, "delay", 0.5
outs a1, a1
```

```
ga1 = 0
endin

</CsInstruments>
</CsoundSynthesizer>
```

7.6 Music Programming in a Multi-Language Environment

The embedding of a language such as Faust in Csound, and indeed of others such as Python and Lua, as well as the embedding of Csound within other systems via its API, places the question of a multi-language environment for music programming at the centre of the ideas expressed in this chapter. We believe that such mix of environments not only fits in the separation of concerns paradigm that is commonplace in systems development [1, 10] but also provides a creative hothouse for Ubiquitous Music.

7.6.1 Separation of Concerns in Ubiquitous Music

We proposed and exemplified the concept of Ubiquitous Music Ecosystems through development of support for creative activities using two high-level languages: Csound and Faust/domain-specific languages such as Csound provide very specialised programming environments. They excel in given tasks and are less useful in others. For instance, we can very easily put together a collection of signal processing units that will modify a given sound, which can be provided via a microphone, or via a disk recording, or synthesised on the fly. It is harder, however, to package these audio processing facilities into a stand-alone application using Csound alone.

In this case, we can wrap our synthesis/processing code inside a program written in a language that is designed for making stand-alone applications. For instance, if we are coding for desktop machines, we can use C/C++, Java or Python and with the help of the Csound API put together a program that will control the audio processing in a suitable way. If we are writing a mobile application, then a similar process applies, using a combination of Java, ObjectiveC, Javascript, or other system language, and Csound. It is possible therefore to use the same signal processing code for a variety of targets. This approach provides a solution for the heterogeneous programming problems that arise when developing for mobile (and web) platforms (see Prologue and Chap. 2).

From a lower-level angle, we can exploit languages that facilitate the implementation of signal processing algorithms, such as Faust, to extend the capabilities of a host system such as Csound. This can be done in a dynamic way, as demonstrated by this chapter, allowing us to modify and tune our processes for a given application. In this scenario, Faust provides a flexible means to program lower-level operations, efficient implementations.

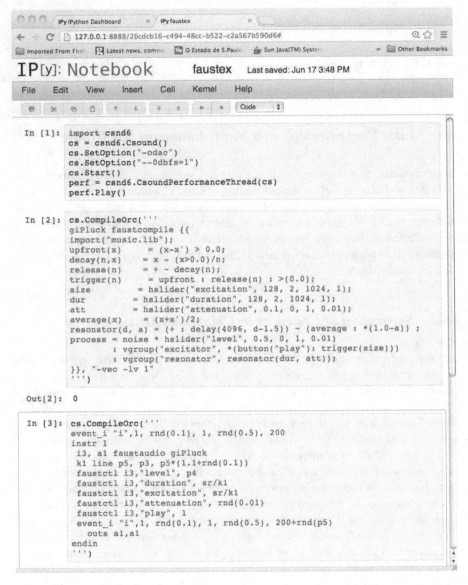

Fig. 7.6 Faust embedded in Csound running under Python in an IPython Notebook

7.6.2 Ubiquitous Music Ecosystems

An example of the concept of Ubiquitous Music Ecosystems (UbiMus ecosystems) is the use of Csound in the IPython Notebook environment, through the Csound API in Python 2.7 (Fig. 7.6). In such environment, which includes a read-evaluate-perform loop (REPL), it is possible to prototype and perform new instruments interactively, with optional user interfaces and through protocols such as MIDI and Open Sound Control. Adding Faust to this combination allows us to write programs and to do sample-level computation at the C language level of performance. Such multi-language, interactive, mix of graphical, controller and text interfaces is an example of a generic, flexible and powerful environment for Ubiquitous Music. This functionality is not achieved by any single language system alone. We propose the concept of Ubiquitous Music Ecosystems for this class of creativity-support environments.

Ubiquitous Music Ecosystems allow users to navigate computer music composition, performance and, improvisation and can be also harnessed for their educational and research potential. Customisation here is a key element, as it is possible to combine the different programming elements in various ways to provide, on one hand, tools for students to learn about sound, music and audio technology, and on the other, environments for cutting-edge research.

Other examples of such multi-language approaches involving Csound include the use of external data-processing systems for algorithmic composition (via its score processor/generator facility); application development for mobile environments such as iOS and Android (also known as the Mobile Csound Platform (MCP) [6]), where Csound is used as the sound engine, while system languages (Objective-C, Java) take care of the UI; and web-based systems [7, 13], where there is an interaction between mark-up (HTML), scripting (Javascript) and in some cases implementation languages (for instance, in the case of NaCl modules [14]; see Chap. 6).

Conclusions

Csound is a domain-specific language designed to describe and control algorithms for sound synthesis and processing. The Csound API allows programmers to create software that uses Csound as their sound engine and to take advantage of the use of multiple languages for Ubiquitous Music applications.

Faust, on the other hand, is a programming language designed to translate flowcharts and implement signal processing algorithms. It can be used to create plugins and components for a variety of environments, including Csound. With Faust 2, the system can be completely embedded in a host. Faust was encapsulated in Csound through a suite of four opcodes that support compiling, instantiating and executing Faust programs as DSP objects. Examples were shown to demonstrate how the host system can be extended with these unit generators.

Multi-language environments are very useful in Ubiquitous Music. They allow the development of customised solutions for different applications, from

(continued)

education to composition, performance and research. The combination of systems discussed in this chapter provides a wide scope for the development of such solutions.

References

1. Damasevicius, R., Stuikys, V.: Separation of concerns in multi-language specifications. Informatica **13**(3), 255–274 (2002)
2. Ffitch, J.: The Design of Csound5. In: LAC2005, pp. 37–41. Zentrum für Kunst und Medientechnologie, Karlsruhe (2005)
3. Keller, D., Flores, L.V., Pimenta, M.S., Capasso, A., Tinajero, P.: Convergent trends toward ubiquitous music. J. New Music Res. **40**(3), 265–276 (2011). doi:10.1080/09298215.2011. 594514
4. Lattner, C., Adve, V.: LLVM: A compilation framework for lifelong program analysis & transformation. In: Proceedings of the 2004 International Symposium on Code Generation and Optimization (CGO'04), Palo Alto, CA (2004)
5. Lazzarini, V., Kleimola, J., Timoney, J., Valimaki, V.: Aspects of second-order feedback am synthesis. In: Proceedings of the International Computer Music Conference (ICMC), Huddersfield, UK (2011)
6. Lazzarini, V., Yi, S., Timoney, J., Keller, D., Pimenta, M.: The mobile Csound platform. In: Proc. Int. Computer Music Conf. 2012, Ljubliuana. International Computer Music Association, San Francisco, CA (2012)
7. Lazzarini, V., Costello, E., Yi, S., Fitch, J.: Csound on the web. In: Proceedings of the Linux Audio Conference (LAC2014) (2014). http://lac.linuxaudio.org/2014/papers/23.pdf
8. Mathews, M., Miller, J.E.: MUSIC IV Programmer's Manual. Bell Telephone Labs (1964)
9. Orlarey, Y., Letz, S., Fober, D.: Automatic parallelization of FAUST code. In: LAC2009. Casa della Musica, Parma (2009)
10. Ousterhout, J.: Scripting: higher-level programming for the 21st century. IEEE Comput. **31**(3), 23–30 (1998)
11. Puckette, M.: Max at seventeen. Comput. Music J. **26**(4), 31–43 (2002)
12. Vercoe, B.: The Csound Reference Manual. MIT, Cambridge (1986)
13. Wyse, L., Subramanian, S.: The viability of the Web browser as a computer music platform. Comput. Music J. **37**(4), 10–23 (2013)
14. Yee, B., Sehr, D., Dardyk, G., Chen, J.B., Muth, R., Ormandy, T., Okasaka, S., Narula, N., Fullagar, N.: Native client: a Sandbox for portable, untrusted x86 native code. In: 2009 IEEE Symposium on Security and Privacy (2009)

Index

© Springer International Publishing Switzerland 2014

D. Keller et al. (eds.), *Ubiquitous Music*, Computational Music Science,
DOI 10.1007/978-3-319-11152-0

Printed in the United States
By Bookmasters